Nuclear Energy Safety and International Cooperation

Twenty-five years after the Chernobyl explosion, disaster struck once again after a tsunami overwhelmed the considerable safety measures at the Fukushima nuclear power plant in Japan. However, Fukushima had in place a solid containment structure to reduce the spread of radiation in the event of a worst-case scenario; Chernobyl did not. These two incidents highlight the importance of such safety measures, which were critically lacking in an entire class of Soviet-designed reactors.

This book examines why five countries operating these dangerous reactors first signed international agreements to close them within a few years, then instead delayed for almost two decades. It looks at how political decision makers weighed the enormous short-term costs of closing those reactors against the long-term benefits of compliance, and how the political instability that dominated post-Communist transitions impacted their choices. The book questions the efficacy of Western governments' efforts to convince their Eastern counterparts of the dangers they faced, and establishes a causal relationship between political stability and compliance behavior. This model will also enable more effective assistance policies in similar situations of political change where decision makers face considerable short-term costs to gain greater future rewards.

This book provides a valuable resource for postgraduate students, academics and policy makers in the fields of nuclear safety, international agreements, and democratization.

Spencer Barrett Meredith, III has worked in the field of international relations as a professor and practitioner for more than a decade, teaching at the undergraduate and graduate levels, as well as serving as a Fulbright Scholar in the Caucasus and a guest lecturer for the US Department of State in South and East Asia. He is currently an Associate Professor at the College of International Security Affairs in the US National Defense University.

Routledge Studies in Environmental Policy

Land and Resource Scarcity
Capitalism, Struggle and Well-being in a World without Fossil Fuels
Edited by Andreas Exner, Peter Fleissner, Lukas Kranzl and Werner Zittel

Nuclear Energy Safety and International Cooperation
Closing the world's most dangerous reactors
Spencer Barrett Meredith, III

Nuclear Energy Safety and International Cooperation

Closing the world's most dangerous reactors

Spencer Barrett Meredith, III

Routledge
Taylor & Francis Group
LONDON AND NEW YORK

earthscan
from Routledge

First published 2015
by Routledge

2 Park Square, Milton Park, Abingdon, Oxon OX14 4RN
711 Third Avenue, New York, NY 10017, USA

Routledge is an imprint of the Taylor & Francis Group, an informa business

First issued in paperback 2016

British Library Cataloguing-in-Publication Data
A catalogue record for this book is available from the British Library

Library of Congress Cataloging-in-Publication Data

Meredith, Spencer Barrett, III.
Nuclear energy safety and international cooperation : closing the world's most dangerous reactors / Spencer Barrett Meredith, III.
pages cm
1. Nuclear power plants—Former Soviet republics—Safety measures—Case studies. 2. Nuclear power plants—Europe, Eastern—Safety measures—Case studies. 3. Nuclear power plants—Safety measures—International cooperation. 4. Nuclear power plants—Safety regulations. 5. Nuclear power plants—Decommissioning—Case studies. I. Title.
TK1362.F6M47 2014
363.17'99526—dc23 2014002180

ISBN: 978-1-138-01850-1 (hbk)
ISBN: 978-1-138-28236-0 (pbk)

Typeset in Times
by Apex CoVantage, LLC

Soli Deo gloria

Contents

Preface

This project began in 1999 as part of my time at the US Department of State during a break from doctoral studies. It included first hand interviews with many of the principle players in nuclear safety in the United States and the European Community. Later research at the International Atomic Energy Agency (IAEA) in 2002 in the midst of Western compliance efforts added the views of Eastern European nuclear officials. The final project came together after Lithuania officially closed the last reactor at the Ignalina power plant in 2009, thereby closing a dangerous period in global environmental safety. Unfortunately two years later, and twenty-five years after Chernobyl exploded, the Fukushima disaster occurred, once again drawing safety concerns to the nuclear industry. This book examines the first efforts to close dangerous reactors, with the hope that the world will need no further reminders of the risks of delaying disaster.

Acknowledgements

I wish to thank my wife for her ongoing support and encouragement through many late nights working. I also thank my sister for her editing of this manuscript from its first, very rough draft, and my father for his sacrifice for me to complete my doctoral degree that began this research. In addition, my children have been a great motivation to do this job well and on time so I can play when the work is done. Finally, I offer my profound gratitude to Carol Kessler, without whom none of this would have been possible.

1 Political rules and nuclear safety

Nuclear power is here to stay. It produces fewer greenhouse gases than fossil fuels, and despite enormous start up costs, is relatively cheap over decades of operation. Unfortunately, nuclear power also generates by-products whose harmful effects are not measured in generations, but in millennia. Even in the United States, the ongoing failure to establish a long-term waste storage facility highlights concerns of future contamination, as well as the potential acquisition of radioactive materials by terrorist organizations. Making matters worse, simply operating nuclear power plants can expose local populations to lethal radiation, as seen after the Fukushima nuclear meltdown in March 2011. The level of damage there surpassed the considerable safety measures at the facility, showing the worst-case potential of continuing global contamination for otherwise "good" reactors.

That disaster occurred twenty-five years after the world witnessed the worst-case scenario for a "bad" reactor when the nuclear power plant at Chernobyl exploded in April 1986,[1] spewing radioactive material over the surrounding countryside and as far away as North America and Japan. Once lush areas of Ukraine and Belarus became uninhabitable, and decades later, Moscow fruit vendors still sold contaminated wild berries collected hundreds of miles away. As bad as that accident was though, Chernobyl also pointed to a series of problems with an entire class of Soviet-designed reactors, problems that could easily have led to similar disasters throughout the region.

The Soviets built nearly seventy reactors during the Cold War, approximately half of them in Russia and the rest going to proxy states in the Eastern bloc and three Soviet republics.[2] The two main types consisted of carbon based, graphite-moderated (RBMK) and pressurized, light-water cooled (VVER) reactors.[3] Graphite-moderated reactors also use water to cool the nuclear reaction, but problems can occur when water turns to steam thereby reducing the water's safety effects on the fissile material, known as the void coefficient. Positive void coefficients are particularly dangerous in RBMKs because they can lead to an increase in the fuel cycle that culminates in core meltdown. Control rods prevent the reaction from getting out of control, but the accident at Chernobyl showed what happens when that control is lost – engineers had been conducting a low-level test that failed to supply enough power to the control rods in time to prevent the disaster. Soviet nuclear officials raised the void coefficient safety standards for all RBMKs after the accident, and after the Soviet Union's collapse a great deal of international assistance went into upgrading the computer systems that control the nuclear reaction. However, fundamental flaws in the RBMK design remained unchanged for decades.

In contrast, VVER's models follow a similar design to Western reactors by using compressed nuclear fuel pellets kept under pressure. Water cools and regulates the reaction, which slows down in the event of a loss of water pressure. The Soviets designed three models based on the size of pipe ruptures that each type could safely sustain before catastrophic failure: 440/213,[4] 440/230, and 1000 (the only new VVER models currently produced by Russia.) The VVER designs offer greater inherent safety measures than the RBMKs, but the older 440/230s and 440/213s lacked uniform safety standards resulting in *ad hoc* procedures across the industry.

Much of this information remained secret during the Cold War until its end revealed the poor condition of all Soviet-designed reactors. At that time, the International Atomic Energy Agency (IAEA) identified numerous problems with all the RBMK and VVER models, claiming that many were too unsafe to continue operating in their present conditions. The list of deficiencies covered everything from poor safety systems and training for plant personnel, to faulty engineering designs and low quality construction materials.[5] Given enough money and opportunity to bring in Western technical skill, upgrades could overcome many of the problems, but nothing would solve the fundamental design flaws in the RBMK and VVER 440 models as neither have external containment structures to prevent the release of radiation in the event of an accident.

The Chernobyl effect

Despite the myriad problems with the Soviet-designed nuclear reactors, the industry as a whole has had strong support since the first commercial units came online in the 1950s, and prior to the 1986 disaster, nuclear power had proven itself a valuable addition to fossil fuels. French nuclear output rose significantly with the construction of nearly fifty reactors from 1963 until Chernobyl's explosion, while the United States followed a similar path despite the accident at Three Mile Island in 1979. Unlike that facility though, Chernobyl lacked the concrete and steel housing that had saved the surrounding Pennsylvania countryside eight years earlier.

Making matters worse, the details of the accident remained a closely guarded Soviet secret, even from those who experienced its immediate effects. Kiev residents received no warning that radiation had reached their city, and while party officials rushed to secure the city's supply of radiation-absorbing medicine for themselves and then remained indoors for several days, the masses were told to participate in the annual May Day parade despite the contamination risks. Moscow kept the event out of the media for several days until international sources detected radiation spikes outside the Soviet borders, and only after the world learned that something terrible had happened did the Soviet leadership finally release information about the incident.[6]

On the fateful day, operators lost control of the nuclear reaction in unit 4, and fires quickly spread throughout the facility. They managed to shut down the remaining three reactors quickly, but could do nothing to prevent the roof from exploding over the damaged reactor. Radioactive material poured out of the gaping hole for ten days, and while most of the material had very short half-lives and dissipated quickly, dangerous levels of iodine, cesium, and strontium penetrated the soil and water table around the plant, while prevailing winds carried lethal materials to other areas of Ukraine, Belarus, and Russia as well. Eventually, the area around the plant became a modern-day ghost town after the governments of Ukraine and Belarus created exclusion zones surrounding the most dangerous hotspots. In the end, casualties included dozens of firefighters and plant personnel who died from acute radiation poisoning; several hundred more were hospitalized, their life expectancies greatly shortened as a result of exposure. The most widespread effects led to a significant rise in leukemia, thyroid cancer among children, and radiation damage to unborn babies resulting in severe mental retardation and cancer.[7]

Chernobyl showed just how bad it could get when reactors lack a failsafe containment structure, yet regrettably, it was not alone in suffering from a litany of serious safety deficiencies, a sobering fact that participants at the 1992 G7 Munich Conference pledged to correct. Western donor states and organizations offered to help those countries in Eastern Europe and the former Soviet Union that operated similarly unsafe reactors by providing technical assistance as part of larger democratization aid packages. Most importantly, the assistance remained strictly conditioned on commitments to close all such first-generation Soviet reactors before the end of their design lives – the estimated length of operation determined prior to construction.[8] The goal was to prevent another Chernobyl from ever happening again, and the governments of Ukraine, Armenia, Lithuania, Slovakia, and Bulgaria agreed. Over the next four years each committed to improve day-to-day operational safety, develop bureaucratic agencies capable of ensuring the success of those efforts, and decommission all reactors lacking containment structures within set timetables. Amazingly though, with the threat of disaster looming ever closer with each day of unsafe operation, those same governments balked at the requirements they themselves had signed, and disasters like Chernobyl remained a realistic possibility for almost two decades.

Since the beginning of Western safety efforts twenty years ago, no study has been done to explain either that resistance or the final decisions to shut down dangerous nuclear reactors. This book fills that gap. The project focuses on countries that have operated first-generation Soviet nuclear reactors and made binding agreements regarding them in order to receive benefits from future cooperation with the West, ranging from trade and diplomatic agreements to full EU membership. Accordingly, each country promised to 1) improve operational safety based on Western technical assessments, 2) create an autonomous regulatory agency with financial and legal authority to implement required upgrades and establish decommissioning protocols, and 3) set aside necessary funds then shut down RBMK and VVER 440 reactors before the end of their design lives. These three conditions serve as the three levels of compliance within each country examined in this study. Soviet centralization produced a common background in training, reactor design, and fuel processing, as well as dependence on Moscow for guidance in resolving operational problems, all of which enables comparisons across the cases as well.

Russia meets part of the criteria but remains outside the study because no Russian government has agreed to the early-closure requirement for international assistance as offered to its former satellites. Nor has any Russian leader had the option of full compliance as a condition for gaining future cooperation with the West, the central benefit to aid recipients in this study. Excluding Russia from a project on nuclear energy seems counterintuitive at first, but the goal of this study is to explain comparable ranges of compliance outcomes, not present an exhaustive list of nuclear power producers and their safety records.[9]

Compliance – Choosing "the devil you don't know"

The end of the Cold War produced a significant increase in compliance studies as the overarching global competition between the superpowers no longer constrained the types and effectiveness of international agreements. Monetary and environmental regimes developed out of the strengthening of European unity and rising international concerns about climate volatility, while academic studies tended to focus on three primary categories of variables in the creation and maintenance of those agreements. The first evaluated the characteristics of those international regimes, such as the durability and uniformity of rules, the types of monitoring procedures for participating governments, and the narrowness of the regime's focus seeking to prevent global catastrophes or more local problems.[10] A second group focused on the nature of incentives offered by foreign governments through "carrots and sticks" and their relevance to domestic actors,[11] while a third approach examined enforcement mechanisms to deter non-compliance either through increased reputational costs (shaming and the loss of future cooperation) or overt penalties (monetary, legal, trade, etc.).[12] Across these studies, constructivist models of socialized learning utilized inter-governmental dialogue and information exchanges, as well as the normative arguments used in the development of international human rights agreements.[13]

These studies generally explained compliance through international factors rather than domestic ones. As such, none could explain variations in compliance with nuclear safety agreements in Eastern Europe and the former Soviet Union because neither regime types, nor incentives, nor enforcement mechanisms varied at the international level. Western assistance was a) comparably generous to each country,[14] b) future benefits consistently depended on compliance, and c) enforcement remained weak compared to other international agreements.

Peering into the domestic level, strategic bargaining models propose that officials may simply want the benefits that go with being a member of the in-group, and therefore promise compliance with little or no intention of carrying out their obligations over time.[15] Signature becomes a way to defray criticisms by international advocates in more powerful countries, or it functions as a means of delaying real change during the interim years between signature and ratification – buying time and buying off opponents remain two possible reasons for signing an international agreement.[16] However, all the countries under examination here complied to some meaningful and costly degree with their agreements. Therefore, whatever the intentions regarding initial signature, something else drove the actual behavior of each country. This book argues that compliance was universally conditioned by elite assessments of the domestic costs and benefits of keeping those promises.[17]

Domestic cost/benefit analysis appears as a background condition in previous compliance studies,[18] and while some have suggested that domestic governmental stability in particular may be a factor affecting compliance behavior,[19] previous models miss the important role that variations play by treating stability as an assumption. In contrast, this study argues that stability of the rules of the domestic political game determined elite decision making regarding nuclear safety compliance. Since these "rules effects" have not been examined in the context of compliance, nor has the compliance literature addressed nuclear safety agreements after the breakup of the Soviet Union,[20] this project presents a necessary addition to studies on international law, Eastern European and post-Soviet regions, energy and environmental policies, and governance research more broadly. It also has a high degree of generalizability to other subjects with comparable costs and benefits.

The Western perspective

In contrast to the model presented here, the leading approach by Western government agencies has focused on the normative safety culture held by nuclear industry and government decision makers, categorizing risk propensity as either averse or tolerant. Western safety standards have been held up as safe (risk averse) with pressure applied to teach Eastern European officials to forsake their unsafe views (risk tolerance) in favor of the stronger Western norms.[21] Each of the international agreements regarding nuclear safety in the region espoused this kind of risk aversion combined with a teaching approach to increase compliance, arguing that acceptance of those norms would enable elites to come to their senses about the risks of operating unsafe nuclear reactors.[22]

This approach explained compliance as normative "Westernization" and noncompliance as a persistence of backwards, unreasonable risk tolerance. Partial compliance did not represent a distinct choice based on evaluations of costs and benefits since elites did not differentiate the costs and benefits of compliance, instead simply viewing one approach as better than the other. Consequently, this model explained the Chernobyl disaster as a result of risk tolerance that allowed fundamental design deficiencies to creep into the region's nuclear power plants. Those flaws demanded drastic measures to correct, and Western agencies interpreted the reluctance of Eastern nuclear and political officials to do so as a foolish belief that their designs were safe enough in their present conditions.

Exemplifying the point, a US Department of Energy official commented to me on the slow progress of nuclear safety work in the region during the years of Western assistance. He wondered with a great deal of frustration, "Don't they realize what they are dealing with? What the hell are they thinking?"[23] His reaction was understandable given the assumption that Eastern European and former Soviet elites would comply if they properly understood the risks. The perceived lack of such an understanding formed the basis of extensive Western education programs in the region.

To that end, the IAEA worked for years to change Eastern European safety culture as a way to increase compliance. Miroslav Lipar, a former Slovak nuclear regulator working as head of the IAEA Operational Safety Section, showed me scores of documents on training conferences, internal workshops, and collaborative efforts with Western nuclear industry groups that all stressed safety culture as the most important aspect of compliance.[24] This approach was not limited to the IAEA though. I also met with several US Environmental Protection Agency and Department of Energy officials who stressed the need to "get inside their heads and make them think the right way" – restating that safety culture must be the priority for compliance.[25] However, the underlying assumptions were incorrect because they misunderstood compliance, which was always more complex than this approach allowed as it varied over time, while the domestic safety culture in the target countries remained almost universally risk tolerant. Even more so, the safety culture model could not adequately explain partial safety improvements short of full decommissioning since fixing some of the operational dangers included an awareness of what could go catastrophically wrong; improvements in one area required later levels of compliance to ensure the worst did not happen. In addition, decreasing compliance posed an even more serious problem for the model by showing that extensive international efforts to change risk propensity were largely ineffective, thereby calling into question the process of norm education. In both instances, something else influenced compliance behavior, namely the political context in which elites made evaluations about the risks of operating their nuclear power plants.[26]

In particular, stability in the rules of the political game conditioned the evaluations of short-term costs and long-term benefits, the heart of nuclear safety compliance in these cases.[27] The lack of such stability can explain why some governments were reluctant to close down reactors that met energy needs, despite their poor safety records. A critical element of that perspective appeared when I asked the Armenian Ambassador to the International Atomic Energy Agency why his government kept failing to keep its commitments. He explained that the government's decisions depended on degrees of risk: the risk of a nuclear accident versus the risk of the population freezing to death.[28] Beyond such practical considerations, he added that his understanding of risk depended heavily on the stability of the political system, whether the system would survive into the foreseeable future and how long he could realistically expect to participate if it did.

How politicians understand risks depends on many factors, and this study focuses on personal experiences in political transitions (notably turmoil in the government, economy, and societal connections), and learning based on the experiences of others under similar conditions.[29] That kind of learning can occur at the individual level, within a community, or both, and the skills come from some combination of normative values, utility calculations, and cognitive mapping.[30] Too often though, insufficient information prevents a completely accurate risk assessment, and decision makers face constraints due to bounded rationality and preconceptions derived from biased information.[31] Long periods of political transition can also increase the likelihood of uncertainty because the degree of stability 1) effects preference ordering; 2) draws greater attention to core cultural values, which may be opposed to norms espoused by international agreements; and 3) can make short-term thinking more important in utility calculations.[32] Once the decision is made though, elites still face the problem of applying the correct information since self-fulfilling prophecies, *ex post facto* rationalizations, and cognitive dissonance can limit the process of effective learning about risks as well.[33]

Government decision makers in Eastern Europe and the former Soviet Union were especially hard-pressed to make accurate risk assessments as they had to overcome severe economic hardships during their transitions from communism, while also trying to balance social pressures for political change. Widespread popular support for peaceful transitions toward democracy and democratic consolidation afterward aided them in this process, and some were quite successful. Even so, most endured major challenges as the outcomes remained uncertain for a long time.

Rules of the political game become even more important in the context of such transitions. They influence notions of community survival and personal involvement in the process, which serve as key elements in risk assessment.[34] They also define the totality of acceptable behavior and power relationships between political decision makers,[35] and the stability of those structures and processes shape how willing political actors will be to pay present costs to gain future rewards.[36] Considerations for nuclear safety compliance were particularly challenging because costs required payment almost exclusively in the short term, while benefits did not accrue for several years to come.[37] This book examines the context for those evaluations, specifically how the stability of the rules defining the political system determined elite decision making regarding their countries' unsafe nuclear reactors.

Causality of the rules of the game

Domestic rules of the game belong to a larger body of literature dealing with stability in different types of political systems. The democratization literature has implicitly touched on aspects of rules stability,[38] but only as part of a broader concern with transitions and democratic consolidation.[39] A recurrent shortcoming exists when rules effects enter the literature because of the assumption that rules must be formalized in a constitution in order to be classified as stable.[40]

Early on, Almond and Coleman (1960) offered a way around that assumption by looking at framework rules that define the exercise of power outside a formal, constitutional structure. They compared governance mechanisms in a variety of systems to show that *governance mechanisms* can function as more than just extensions of *government institutions*. As an example, clan patronage may define the real practice of politics, but so can official debates set within formal institutions.[41]

All of the countries in this study operate in the framework of *formal* constitutions that define the structure of government. The model used here can also apply to countries without codified constitutional frameworks since alternative rules could include dynastic succession or religious appointment for example, so long as the governing authority determines, establishes, and maintains the formalized rules. In addition, *informal* rules represent the modus operandi of daily governance. These can exist within government structures as party systems, bureaucracies, or legislative committees, although traditional patronage relationships that pre-date or supplant formal institutions may also contain informal rules. Clan identities, corruption, and organized crime can also function as non-institutional processes if the real practice of politics takes place through them.

Formal rules exist in government structures and informal rules in the normal processes of day-to-day governance. These a) define and delineate power between areas of government, and b) establish procedures for determining who is in office and for how long; the *clarity* of the delineation of power and the *predictability* for the selection process defines the stability of the rules. Employing categories of "clear/unclear" and "predictable/unpredictable" gives effective measures of the formal rules seen in official constitutions, amendments, relevant legislation, and presidential decrees; and for informal rules found in cabinet assignments, government turnover frequencies (pre-term, term, post-term), and ministry realignments and expenditures.[42]

Information for short-, medium-, and long-term assessments of rules stability in this study comes from frequent reviews in the *East European Constitutional Review* (*EECR*), cross-country studies examining post-communist transitions and democratic consolidations,[43] and annual governance data from Transparency International, Freedom House, and the *Nations in Transit* (*NIT*) series during the periods of compliance.[44] In particular, *NIT* offers broad evaluations of democratization, economic liberalization, and the rule of law, and makes annual comparisons of each; rule of law scores provide the most comprehensive corroborating evidence for the detailed empirical sources.[45] These three sources make it possible to assess rules stability fairly accurately by showing variation across and within cases.

The following table shows several examples of different elements within each rule type used in this study. Neither set of rules has more inherent causal significance to compliance behavior than the others, nor do individual elements within each type of rule carry greater relative weight.[46] The table focuses on the overall stability of each rule type and the combined impact on the political system.[47]

Table 1.1 Rules of the Game

Rule Type	*Power Considerations*	*Selection Processes*
Formal	• Legal separation of powers between branches of government • Presence of overlapping authority and mechanisms for resolving conflicts of interest • Legislative authority vis-à-vis presidential decrees • Location of governmental power: to whom is the Prime Minister beholden? • Presence of political immunity	• Election laws • Impeachment laws • Parliamentary dismissal procedures • How the Prime Minister and government are nominated, approved, and dismissed
Informal	• Bureaucratic responsibilities, presence of overlapping authority, and conflict resolution mechanisms • Ministry positions in government hierarchies • Realignments to different ministries • Ministry autonomy as stipulated in the formal rules vs. actual practice	• Presence of competing patronage hierarchies • Procedures for "moving up" • Public/private dismissal practices

Applying these variables to three categories, first, the presence of unclear power considerations and unpredictable selection processes in both sets of rules results in a ranking of **low stability**. Second, **medium stability** exists when only one set is clear and predictable, and third, clarity and predictability in both types represent **high stability**. These three categories capture a wide range of rules effects and produce greater explanatory accuracy than simply labeling the rules as stable or unstable. The following breakdown gives more information about each level of rules stability.

- *High Stability*: Formal and informal rules are clear and predictable, enabling certainty about the longevity and resilience of those rules to weather diverse international and domestic crises. Evidence for stable rules resides in the predictable transfer of power through regular elections, hereditary title, selection by clan leaders, or some other reliable method that continues over time. Stable rules also exist in a clear differentiation of authority between groups in the political system and a lack of challenge to said delegation.

Table 1.2 High Stability

Rule Type	*Clarity*	*Predictability*
Formal	Yes	Yes
Informal	Yes	Yes

- *Medium Stability*: This condition results from inconsistent levels of stability between formal and informal rules – one set is stable while the other is not.[48] Stable governmental structures may exist with unstable processes of day-to-day governance, or a condition of stable informal rules and unstable formal rules could be seen in transitional political systems that operate on patronage relationships while elites try to work out the institutional structures of government. Medium stability also describes key periods of transition to either endpoint. Movement toward increasing stability may be hindered by lingering threats to the developing political order, or declining stability can be held in check by certain accepted practices that prevent the slide from going all the way to the bottom. Each element of rules stability – clarity and predictability – can also vary across rules types. The following tables show the possible conditions for medium stability.

Table 1.3 Medium Stability (a)

Rule Type	*Clarity*	*Predictability*
Formal	Yes	Yes
Informal	No	No

Table 1.4 Medium Stability (b)

Rule Type	*Clarity*	*Predictability*
Formal	No	No
Informal	Yes	Yes

Table 1.5 Medium Stability (c)

Rule Type	*Clarity*	*Predictability*
Formal	Yes	No
Informal	Yes	No

Table 1.6 Medium Stability (d)

Rule Type	*Clarity*	*Predictability*
Formal	No	Yes
Informal	No	Yes

Table 1.7 Medium Stability (e)

Rule Type	*Clarity*	*Predictability*
Formal	Yes	No
Informal	No	Yes

Table 1.8 Medium Stability (f)

Rule Type	*Clarity*	*Predictability*
Formal	No	Yes
Informal	Yes	No

- *Low Stability*: Both types of rules are characterized by unpredictable methods of power transfer and unclear delegation of political authority, with persistent challenges to that delegation defining the political system. The structures of government and the processes of governance remain unresolved and impermanent. Arbitrary decision making and unpredictable changes to ministerial and governmental responsibilities produce uncertainty about how actors should behave in the present, while challenges to the formal delegation of power in government structures have the same negative effect.

Table 1.9 Low Stability

Rule Type	*Clarity*	*Predictability*
Formal	No	No
Informal	No	No

The rules effects model used here predicts that high stability should be the least common condition in the former Soviet/Eastern bloc countries during the periods of their compliance behavior because of the significant governmental, economic, and societal transitions experienced in that time. Medium stability should occur most frequently because it captures the contradictory tendencies for some elements to stabilize while others lag behind. Finally, divisions between competing elites should also be more moderate than in openly contested systems under low stability, which resemble the "Wild West" in their arbitrariness and recourse to non-legal means for conflict resolution in the political arena.

All three types of stability shape elite expectations about the future since they share a common concern for the survival of the political system, even if the likelihood of their personal participation in it matters more immediately.[49] Those concerns shape elite willingness to endure immediate costs in order to gain future rewards.[50] The relative size of costs and benefits are key elements of elite decision making, but expectations about the long-term viability of the political game and their participation in it provide an overarching framework in which cost/benefit analyses occur.[51] The following diagram represents that process.

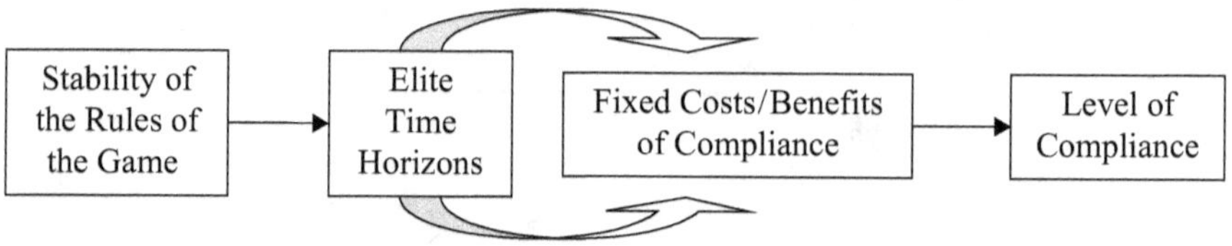

Figure 1.1 Rules Effects Causal Mechanism

This approach assumes a rational model for elite evaluations of costs and benefits, as well as for assessments of the clarity and predictability of political rules. Direct cognitive assessment was not always possible given limited access to some decision makers, and the possibility of instrumentality in interviews remained a methodological hurdle when granted access.[52] Inference of rationality offers a more practical approach given two assumptions: one can expect actors to perceive events reasonably accurately, even if risk assessment and learning are not perfectly rational, and anyone in similar situations with similar learning and experiences would produce the same basic evaluations.[53] This allows for a determination of probabilistic causality, the best one could hope for in a longitudinal cross-country study.

Given this assumption of common, rational elite perceptions of the durability of political rules, it follows that their utility calculations of short- and long-term costs and benefits were in turn affected by considerations of rules stability.[54] Accordingly, low stability created disincentives to pay short-term costs to gain long-term benefits, high stability reversed the incentive structure, and medium stability created contrasting incentives leading to a combination of short- and long-term goal seeking. This impacted compliance with nuclear safety agreements due to their immediate costs and long-term benefits.

Immediate costs varied in part based on dependence on nuclear energy for domestic consumption and electricity exports.[55] Each of the five countries in this study has relied heavily on nuclear power for energy production, ranging from seventy-eight percent for Lithuania to forty percent in Armenia. Costs have gone beyond revenue concerns though as elites faced political costs for compliance as well.[56] Even Chernobyl, with its partial meltdown in 1982 and catastrophic failure four years later, and Metsamor, situated on an active seismic fault in Armenia, were used as popular rallying cries against foreign intervention; the less obviously dangerous reactors even more so.

Each country has also struggled to meet basic social welfare needs while struggling with a host of energy-related problems: transitioning to alternative energy sources, increasing production efficiency,[57] implementing safety upgrades and new training methods with considerable Western assistance, developing additional layers of bureaucracy for an independent nuclear regulatory authority, and setting aside vital financial resources for decommissioning. While countries have varied in their dependence on nuclear power, the range of compliance expenses was universally staggering, and unfortunately occurred largely as up-front costs. Closing Chernobyl cost nearly $1 billion for example; compliance with all the other areas at the plant cost several hundred million more. Long-term benefits offered a net gain by comparison, but it was not always easy for elites to accept them as such;[58] even the daily value of increased safety accrues over time making it primarily a long-term benefit. Reputational benefits were also weighted toward future interactions, and cooperation with the West, whether through official EU membership or beneficial financing and infrastructural grants, pays its greatest rewards over long periods of time. The net result of elite evaluations of these costs and benefits was slower compliance than the West hoped, but still much faster than many in the East wanted.

Nuclear safety compliance

Simply operating unsafe reactors presents considerable risks, and the approach used here challenges existing explanations in the compliance literature and among Western governments to show why elites chose not to take the necessary steps to eliminate those risks. It does so by stepping outside the common technical debates about standards and design lifetimes that hindered safety work,[59] and outside the debates about international regime characteristics by emphasizing domestic receptivity to international offers rather than the quality of those efforts themselves.[60]

Like the rules shaping receptivity, compliance behavior shows more nuance than revealed by a dichotomous "compliance/noncompliance" approach. A more varied approach has merit given that all the cases exhibit at least a measure of partial compliance.[61] In addition, the presence of three common requirements for compliance, ranked according to their relative costs, makes it possible to use "low/medium/high" values for compliance, comparable to the levels of rules stability. In particular, the lowest level occurred with *safety modifications* to technical components, personnel training, and operational procedures; medium costs increased with *regulatory sufficiency* in finances, bureaucratic autonomy, and control of the necessary personnel and equipment to ensure operational safety until decommissioning; and finally, the highest took place with *early closure* of unsafe Soviet-designed nuclear reactors, including the necessary precursory step of establishing and funding decommissioning programs beforehand.

These costs extended beyond financial resources so that each level became progressively more expensive politically as well; regulatory sufficiency cost more than upgrades because it entailed changing existing laws, reallocating budgets with the concomitant need to reach political consensus, paying salaries, establishing new infrastructure, and training appropriate personnel. Compliance in this area also opened up the possibility of allowing the regulator to shut down the plant for either short periods of time for safety reasons, or force long-term closure due to fundamental design deficiencies. Decommissioning posed the highest costs because it included all of the above costs, as well as the allocation of funds to pay for future decommissioning, development and integration of alternative energy sources to replace the defunct reactors, and the considerable costs of safely storing all radioactive material.

Defining compliance categories this way allows for consideration of both the quality and the quantity of compliance decisions. Therefore, levels are progressive and cumulative so that the highest level of compliance can only occur after the preceding two conditions have been met.

Table 1.10 Levels of Compliance

Low Compliance	Safety upgrades		
Medium Compliance	Safety upgrades	Regulatory sufficiency	
High Compliance	Safety upgrades	Regulatory sufficiency	Reactor shutdown and necessary precursory steps

Requirements for each level of compliance come from multiple sources including the Convention on Nuclear Safety (CNS),[62] various Grant Agreements funded by the European Bank for Reconstruction and Development (EBRD),[63] and specific Memoranda of Understanding (MoU) with the European Commission and the United States. Short-, medium-, and long-range sources also give an accurate picture of the level of compliance in each country. These include 1) industry journal articles; 2) independent assessments by non-governmental organizations, like Bellona and Friends of the Earth; 3) annual reviews by the IAEA, the EBRD, the Nuclear Energy Agency of the Organization for Economic Cooperation and Development (OECD-NEA), the European Council (EC), and the International Nuclear Safety Program of the US Department of Energy; and 4) interviews with Western experts in the OECD-NEA, the US government, the IAEA, and Foratom – the Western European nuclear lobby to the European Union.[64] Interviews with nuclear regulators, plant managers, and representatives from nuclear energy ministries and design institutes for each of the complying countries complete the evidentiary requirements of the study.[65]

Afterword

This book argues that the level of stability in the political rules of the game explains compliance better than the Western safety culture model, despite its preeminence in policy circles and among Western nuclear officials. It does so as more than an academic debate though – nuclear reactors pose tremendous risks if designed or operated unsafely; Fukushima showed what can happen at even the safest facility. As such, the topic matters for ongoing discussions about nuclear power as the industry grapples with its future, and governments strive to meet public needs for safe energy. The model also has generalizability for other policy areas, the subject of a forthcoming study on democratization under conditions of conflict.

It is also more than "ivory-tower" analysis, removed from the day-to-day struggles experienced by the people in the region. The book strives to bring to life the human side of hard choices and complex decision making in the midst of freezing winters, fist fights in the halls of power, widespread corruption, and murder for political gain. Countering this dark side of nuclear safety compliance have been the many examples of personal sacrifice, professional integrity, and dedication to the public good. In that regard, this work is a testament and tribute to those who died fighting Chernobyl, as well as those who worked for decades to prevent, rather than simply delay another disaster.

Notes

1 I have used the Russian spelling for Chernobyl as it has more common usage and acceptance internationally. I have also used transliterations for abbreviations of political parties and organizations whenever possible to avoid the appearance of English-bias, while translating the party names themselves into English to make reading by an English-speaking audience easier. I also use European Community, Commission, Council, and Union appropriately based on the timing of the Union's development and the relevant governing structure indicated in the book.

2 Bulgaria, Hungary, the Czech Republic, and Slovakia in Eastern Europe; Armenia, Lithuania, Ukraine, and Kazakhstan in the former Soviet Union.

3 Besides these two main designs, the Soviets also built a few light water, graphite moderated, and fast-breeder reactors in Beloyarsk and Bilibino, but they generated a small percentage of electricity compared to the other main types of Soviet reactors. Kazakhstan had a small BN-350 breeder reactor to power a desalination plant for the capital, but it has since closed.

4 Measurements in millimeters.

5 A compilation of the problems included: inadequate fire protection systems (all), not enough separation between safety systems and the fuel cycle (all), lack of redundancy in plant safety systems (all), badly managed operating and emergency procedures (all), insufficient training methods (all), weak systems for monitoring critical safety parameters (all), limited and ineffective accident mitigation systems (RBMK), inadequate reactor confinement to accommodate large-pipe breaks (VVER 440/230), weak emergency core cooling systems (VVER 440/230), insufficient leak-tightness for pipe fittings (VVER 440/230 and 213), and poor hydrogen mitigation in reactor confinement (VVER 440/230 and 213).

6 Marples (1986) and Medvedev (1991) provided some of the first detailed accounts.

7 Vargo (2000) provides comprehensive data.

8 The term "design lifetime" remains contentious as some Russian and Eastern European nuclear officials argue that initial assessments of reactor functionality at the time of construction were by nature limited and premature given on-the-job learning. Accordingly, plant personnel and regulators should determine how long a reactor remains online, even if that means indefinitely with repairs, replacement parts, and refueling. Many efforts at lifetime extension have been implemented along that logic.

9 The Czech Republic, Hungary, Slovenia, Romania, Finland, Cuba, Iran, India, and China have no place here either. While the Czechs have incurred sizable costs and considerable international opposition regarding their hybrid-Temelin reactor, a VVER 1000 Soviet

reactor that underwent a core redesign to make it more compatible with Western technology, neither Temelin nor the second generation Dukovany plant have been subject to early closure commitments. The same holds for the Paks plant in Hungary. Slovenia and Romania relied heavily on Western design assistance, eliminating the need for early closure, and most of the safety work done in Finland came under the Finns' own initiative. Nor have any of these countries entered into early closure agreements. Kazakhstan could have been included as an interesting case, but its Aktau nuclear reactor does not fit into the same category as the facilities under consideration in the study; there were very low immediate costs to shutting down the reactor since it only provided power for a desalination plant that was easily converted to fossil-fuel produced electricity. Kazakhstan also does not possess comparable potential for long-term benefits from compliance, as do the other countries. Finally, Iran, China, and India developed the safer VVER 1000 reactors, and Cuba never completed the Jaguar plant.

10 Kolben (2010); Vezirgiannidou (2009); Kirk (2008); Sachs (2008); Simmons (2001, 1998); Wettestad (2001, 1999); McGillivray (2000); Underal (1998); and Cowhey (1993) offer excellent examples.

11 See Skerl (2011); Clarke (2010); Mushkat (2009); Mitchell and Hensel (2007); Carrubba (2005); Sverdrup (2004); Knox (2001); Simmons (2000); Victor, Raustilia, and Skolnikoff (1998); and Koh (1997).

12 Wettestad (2001) and Cowhey (1993) present two solid studies. Expanding the field of inquiry, Simmons (2001, 2000, 1998); Downs, Rocke, and Barsoom (1996); and Chayes (1995) address efforts to increase compliance through various methods of international pressure. Managerial regimes differ from coercive ones, which in turn counter more normative approaches.

13 Zartner and Ramos (2011); Ambrosetti (2010); Cardenas (2010); and Risse, Ropp, and Sikkink (1999) give overviews of the larger debates. These contrast legal approaches emphasizing codification and statute clarity. See Johns (2009) for an historical review of those efforts during the first decade of the twentieth century.

14 Assistance types and amounts varied, but each country received similar aid packages according to energy consumption, the presence of surplus energy exports, and alternative energy sources.

15 See Davies (2010) for recent work on rhetorical action through UN mechanisms, while previously Victor, Raustilia, and Skolnikoff (1998) test this nonobservance behavior referencing Soviet and Japanese adoption of and later withdrawal from the International Convention to Regulate Whaling, and Soviet/Russian signature of the London Convention on Nuclear Dumping. See also Fearon (1998) for more on the intentionality of signature.

16 Chayes (1993) and Young (1979) argue that signature reveals an implicit expectation to comply. Given the costs incurred during negotiations preceding the signing of an international agreement, and the assumption that rational actors satisfice their priorities, they maintain that decision makers will initially try to comply. States will not sign without some measure of initial willingness to abide by the treaty, and later divergence is a separate issue. Cardenas (2010) and Risse, Ropp, and Sikkink (1999) provide solid evidence to that effect, and Jönsson and Tallberg (1998) look at ways to improve compliance after the fact. However, Finnemore and Sikkink (1998) also stress the instrumentality of signing agreements. Decision makers may simply wish to alleviate immediate external or internal pressure with no real intention of behavioral realignment. Other approaches rely on the conflicting pressures of two-level game negotiations (Putnam, 1988), or the process of agreement formation (Abbott et al 2000) to remain neutral about intentions, and instead emphasize the constraints under which intentions are processed and articulated.

17 Aside from the intentions behind initial signature or the nature of the agreements themselves, Underal (1998) accurately identifies three general factors affecting compliance outcomes: 1) rational interest calculations, 2) domestic politics, and 3) ideational

structures. He admittedly refrains from testing the causal mechanisms he presents with either empirical evidence or broader theoretical adjudication. In contrast, Weiss and Jacobson (1998) and Mitchell (1994, 1988) broadly test models of international regime construction designed to induce elites toward acceptable behavior, but their work lacks sufficient theoretical grounding. This shortcoming limits the generalizability of their extensive empirical research.

18 For example, Powell and Stanton (2009) evaluate domestic judicial factors affecting signature and compliance, but fail to include the larger domestic factors evaluated in this study.

19 See Checkel (2001c, 1999); Checkel and Moravcsik (2001); and Schimmelfennig (2001) for initial attempts.

20 Foss (1999) looks at nuclear energy and policy making in Eastern Europe and Russia, but as an overview of the technical conditions and international efforts, while Tejada (2005) focuses only on Bulgarian nuclear production and democratic development. Tejada offers the most comparable subject matter to this book. However, he argues that EU incentives did not compensate for undemocratic dealings with Bulgaria, a topic that I set aside, instead holding to the nature of the incentives and their consistency over time rather than evaluating the merits of those efforts. Tejada also suggests that further comparison between Bulgaria, Slovakia, and Lithuania in the areas of compliance and democratization would benefit the subject, something this book addresses in depth.

21 Legro (1997) offers a good discussion on strong vs. weak norms and Finnemore and Sikkink (1998) identify different norm internalization processes. For more general discussions of normative influences see Cortell and Davis (2000, 1996).

22 The tactic has been the basis of IAEA technical exchanges and training programs. While education may improve safety culture, it has not done so with compliance because it does not take into account power relationships between nuclear regulators and political officials or intra-elite competition.

23 Personal Interviews, May 1999.

24 Personal Interviews, September 2002.

25 Wilpert and Fahlbruch (2002) address the gap between attempts to instruct and even coerce better safety approaches with persistent safety challenges. The methods they advocate can improve safety but only within the larger context of how nuclear officials and their political decision makers view the costs and benefits of those measures. Safety culture does not exist in a vacuum.

26 Douglas and Wildavsky (1982) offer an interesting analysis of the mechanism by which norms frame rational evaluations of costs and benefits. For broader discussions of the debate between normative and rational models see Checkel (2001a, b, c, 2000, 1999); Risse (2000);Thomas (2000); Risse, Ropp, and Sikkink (1999); Finnemore and Sikkink (1998); Ruggie (1998); Keck and Sikkink (1997); Legro (1997); Finnemore (1996); Katzenstein (1996); Risse-Kappen (1995); and Goldstein and Keohane (1993).

27 Streich and Levy (2007) delve deeply into the discount rate model to show a broad range of choices that incentivize compliance as cooperation in some areas, but not in others. Their "opening of the black box" coincides with Krebs and Rapport's (2012) work on individual cognitive evaluation of near and distant threats. The inclusion of Construal Level Theory to the compliance literature would add much needed nuance to the field in general, something this book begins to do by looking at nuclear safety in particular. Their work also shows that discounting the future can lead to action rather than inaction more often than traditional models propose; this too fits well with the research and conclusions presented here as the book uses an operationalization of compliance to include more than simply refusing decommissioning by actively seeking lifetime extension.

28 Personal Interviews, September 2002.

29 See Dionne (2011) for the application of elite time horizons to AIDS policy making, and Barkin (2006) for an interesting treatment of decision-making meta-narratives

concerning environmental policy cost/benefit analyses. Botcheva (2001) also offers an excellent look at learning models and the legitimacy of experts in international negotiations, the agreements that follow, and their implementation over time. This issue played itself out often in the countries examined in this book.

30 The complexity of formal risk assessment modeling is often too much for untrained decision makers (Paton, Smith, and Johnston, 2000). The tools used by professional soothsayers require years of study and practice, something politicians and few rank-and-file citizens can afford to pursue.

31 Hertwig, Pachur, and Kurzenhauser (2005) and Greenberg and Schneider (1995) examine mechanisms for evaluating comparable risks, influences on assessment, and the implications for those evaluations on policy decisions.

32 As per Collinson and Holden (2005) and Regan et al (2005).

33 See Brown, Treviño and, Harrison (2005). In addition, state-society relationships play a role with the possibility of elites restricting vital information to the public. This becomes more difficult when risks are obvious and knowledge widely dispersed. Intermediary groups can effectively moderate state risk taking if ample political space and civic commitment for them to operate exist. Japan's nuclear industry provides an example of good information flows, effective NGO representation, and reliable technologies to mitigate the risks of nuclear power (Pickett, 2002), all of which Fukushima tested after the disaster. The disaster at Chernobyl is more complicated, making it an excellent benchmark for precautionary regulatory efforts in Eastern Europe. See Goklany (2001) for more on the use of precautionary principles by environmental groups.

34 Hiskes (1998) addressed multiple social concepts of risk and the ways policy makers engage and shape them.

35 See Acemoglu and Robinson (2001) for an excellent evaluation of relative costs to elites and pro-democratization populace groups during transitions; Anderson, Fish, Hanson, and Roeder (2001) for assessments in the post-communist context; and Eckstein's (1992) seminal work on transitional polities.

36 LaPuente and Nistotskaya (2009) evaluate comparable processes in merit-based bureaucratic governance.

37 Viklund (2004) examines the relationship between public opinion about those costs and benefits and political decision making. While the study is limited to Sweden, with high civic awareness of environmental issues, the general principles of popular inputs over direct costs/benefits also apply to transitioning polities. Etner, Jeleva, and Spaeter (2009) consider risk perceptions as central to those analyses, an approach that can include the connections to nuclear power as powerful popular symbols of national pride as well.

38 Carlin and Smith (2010) and Anderson, Fish, Hanson, and Roeder (2001) effectively review the literature regarding alternative indicators of regime stability.

39 Denk and Silander (2012), Burnell and Schlumberger (2010), Gill (2002), Alexander (2001), Pridham and Agh (2001), Diamond (1999), Nagle and Mahr (1999), Dawisha and Parrott (1997a, b, c, and d), Linz and Stepan (1996), and Przeworski (1995) provide comprehensive treatment.

40 Hale (2011) convincingly shows the opposite for the former Soviet states as informal rules dominate formal constitutions.

41 See Gorenburg (2012); Naurin and Rasmussen (2011); and Alley (2010) for three case studies of domestic and multinational formal rules. See also White (2010) and Finn (2004) for two examples of the debates surrounding the centrality of the rule of law as an indicator and predictor of regime stability.

42 These are outside evaluations based on verifiable evidence. Elite perceptions of the evidence will be addressed later.

43 These give greater context than the more limited focus of *EECR*.

44 Jaggers and Gurr (1996) are also helpful, but their study ends in 1994, too soon to be of much immediate value. *NIT* addresses most of the indicators they examine, thereby providing continuity over the period of this study.

45 Sometimes the authors of *NIT* realign or add categories to their analyses. While any alteration in coding can affect the reliability of the results, *NIT* researchers do so to increase the comparability of annual analyses. However, one potential limitation for using *NIT* occurs with the range of variations between values, which does not correspond identically to my scale. Changes of more than .25 in value indicate a significant outcome in *NIT*, but that scale would place my evidence squarely in the middle with huge ranges at the endpoints. I limit the problem by using *NIT* to evaluate the general findings of the other two sources.

46 Helmke and Levitsky (2004) offer an excellent typology comparing weight of formal and informal rules, as well as the causes for stability in the latter. This book does not necessarily examine causes for instability, rather it operationalizes stability and shows its effects.

47 I do not explicitly identify or evaluate factors contributing to stability conditions. Most of the post-communist literature deals with it in some fashion or another. More work can be done on the relative weight between rules types and the elements within each set, but I willingly sacrifice some explanatory power within each country to gain it across all countries.

48 I do not prioritize one type of rule over the other; stable formal rules will not automatically compensate for the uncertainty resulting from unstable informal rules, or vice versa. What matters is that one is stable and the other is not.

49 Ken Oye (1986) shows the connection between political stability and perceptions from a game theoretic approach. Barkin (2004) also offers an excellent analysis of international negotiating strategies based on time horizons, a similar approach followed in this study.

50 Axelrod (1984) shows that time horizons (discount rates) affect actors' utility calculations for present decisions. If the future looks promising, in that short-term losses can be made up down the road, actors are more willing to pay costs in the present. The discount rate is central to cooperation – high discount rates (short time horizons based on little chance of future rewards for present sacrifices) make it unlikely, whereas low discount rates (long time horizons) enable actors to see beyond present needs and pay upfront costs to gain the possibility of greater rewards down the road. The nature of costs and benefits, temporally and in the relative value of each (whether costs outweigh benefits or vice versa) for actors' utility calculation matter as well. The cases in this study have fixed costs and benefits: costs are high in the short term, and benefits are also high but delayed until sometime in the future (far enough off that actors will not receive the benefits at the same time as they pay short-term costs, but not so distant that they have no hope of enjoying them later). This key factor allows the use of discount rates as a part of the decision-making process elites use when evaluating their international compliance commitments.

51 Elites must consider the durability of the political rules before considering their involvement, a process they may not intentionally engage, but one that implicitly occurs.

52 See Princen (2009) and Patton (1990) for treatment of this in environmental policy decision making.

53 Personality nuances and bounded rationality may play a role, but this approach sufficiently explains the results without recourse to them in the analytical framework. See Shelton (2003) for normative influences, and Herrmann and Shannon (2000) for the psychology of normative (re)evaluation. Payne (2001), Schimmelfennig (2001), and Risse (2000) also discuss learning processes, addressing both rational and normative elements.

54 Geddes (1994); Levi (1988); and Ames (1987) offer excellent analyses of this in unstable political systems.

55 The capacity of the state to comply also factors into the costs of compliance. The ability to collect electricity payments can affect compliance in three areas: paying nuclear power plant and regulatory agency wages, financing safety upgrades, and establishing decommissioning programs. For discussions of the impact of state capacity on elite

behavior see Cirtautas (1995); Evans (1995); Geddes (1994); Hall and Ikenberry (1989); Skocpol (1985); and Mann (1984).

56 Political costs include changing existing policies and increasing interest group conflict, social mobilization against the prescribed behavior, and the personal costs individual elites may pay for being perceived as lackeys of foreign powers.

57 Methods for this have included making generating facilities more competitive, improving commercial and residential insulation, decentralizing heating control so users can determine their own temperature settings, incentivizing increasing payments for electricity bills, and strengthening the transmission infrastructure along the entire electricity grid.

58 The same fixed temporal condition for costs and benefits can be seen in compliance agreements dealing with human rights, fiscal austerity measures, and nuclear disarmament; see Alm, Martinez-Vazquez, and Torgler (2010); Callioni (2008); Simmons (2000); Risse, Ropp, and Sikkink (1999); Chayes (1995); and Nye (1987).

59 The lack of a universally accepted standard for determining sufficient reactor safety severely limits the applicability of a normative "safety culture" approach. Further complicating the matter is the lack of consensus on which body has the authority to apply its particular definition to the industry. The Euratom charter argues that the responsibility lies with individual countries, the European Commission bases its decisions on the G7's containment/no-containment criteria, and IAEA guidelines have not always spelled out the tipping point when a reactor could be labeled safe

60 Thankfully, elites across the cases did not resort to the ultimate partisan exit strategy by claiming some previous government had signed agreements that the new government need not follow.

61 Cardenas (2010) establishes that even when missing the mark fully, symbolic gestures of compliance, like replacing bureaucratic elites who have become undesirable to the West, show more than simply noncompliant behavior. For example, she incorporates lip-service into a continuum of compliance instead of viewing it as an absolute failure to change state behavior. While symbolic changes do not have the same level of compliance as total behavioral realignment, Cardenas shows they have a measure of compliance and this must be viewed as significant. Comparably, this study's nuanced approach to compliance captures a more complete range of outcomes than previous models.

62 The contributors to Leeland (2013) review the status of CNS participants in greater detail than here as the CNS covers far more than simply regulatory sufficiency. The excellent reference book addresses the critical area of waste management as well.

63 I extend my gratitude to Vince Novak, Director for Nuclear Safety at the EBRD, for his generous assistance.

64 All interviews followed a semi-structured format to facilitate common assessment while allowing for a broad range of responses from the participants. Selected interviewees are listed as follows with their titles at the times of interviews and any later ones known to the author: Carol Kessler, former Senior US Coordinator for Nuclear Safety, former Deputy Director General OECD-NEA, former Director Center for Global Security-Pacific Northwest National Laboratory, Chair of the Nonproliferation and National Security Department, Brookhaven National Laboratory; Dick Stratford, Deputy Assistant Secretary of State, US State Department; Karen Henderson, Assistant for International Programs – Office of the Chairman, US Nuclear Regulatory Commission; Tobjorn Norendal, Norwegian Ministry of Foreign Affairs and Chairman, IAEA Contact Expert Group; Jan Stuller, Regional Coordinator Department of Technical Cooperation, IAEA; David Sycamore, Director International Affairs, Foratom.

65 Jozef Misak, former Chairman Slovak Nuclear Regulatory Authority; Miroslav Lipar, former Chairman Slovak Nuclear Regulatory Authority; Yanko Yanev, former Chairman Bulgarian State Committee for the Utilization of Atomic Energy for Peaceful Purposes (CUAEPP); Jivan Tabibian, Armenian Ambassador and Chief of Mission to the Organization for Security and Cooperation in Europe; Ashot Martirosian, Chairman,

Armenian Nuclear Regulatory Authority; Saulius Kutas, Chairman, Lithuanian State Nuclear Power Safety Inspectorate (VATESI); Armen Abagyan, Vice President, Rosenergoatom (Russian State Atomic Energy Utility); Mikhail Khorochev, Nuclear Engineer, Nuclear Power Technology Development Section, IAEA; Yuri Vishnevsky, Chairman, Gozatomnadzor (Russian State Nuclear Regulatory Agency); Vadym Gryschenko, Chairman, State Nuclear Regulator Committee of Ukraine.

References

Abbott, Kenneth, Robert Keohane, Andrew Moravcsik, Anne-Marie Slaughter, and Duncan Snidal. "The Concept of Legalization." *International Organization* 54, no. 3 (Summer 2000): 401–419.

Acemoglu, Daron and James Robinson. "A Theory of Political Transitions." *American Economics Review* 91, no. 4 (September 2001): 938–963.

Alexander, Gerard. *The Sources of Democratic Consolidation*. Ithaca: Cornell Press, 2001.

Alley, April. "The Rules of the Game: Unpacking Patronage Politics in Yemen." *Middle East Journal* 64, no. 3 (Summer 2010): 385–409.

Alm, James, Jorge Martinez-Vazquez, and Benno Torgler, eds., *Developing Alternative Frameworks for Explaining Tax Compliance*. New York: Routledge, 2010.

Almond, Gabriel and James Coleman, eds. *The Politics of the Developing Areas*. Princeton: Princeton University Press, 1960.

Ambrosetti, David. "Beyond the 'Norm Entrepreneur' Model: Rwanda, Darfur, and Social Sanction among UN Diplomats." *Global Society* 24, no. 2 (April 2010): 151–171.

Ames, Barry. *Political Survival: Politicians and Public Policy in Latin America*. Berkeley: University of California Press, 1987.

Anderson, Richard, Steven Fish, Stephen Hanson, and Philip Roeder. *Postcommunism and the Theory of Democracy*. Princeton: Princeton University Press, 2001.

Axelrod, Robert. *The Evolution of Cooperation*. New York: Basic Books, 1984.

Barkin, Samuel, J. "Time Horizons and Multilateral Enforcement in International Cooperation." *International Studies Quarterly* 48 (2004): 363–382.

——. "Discounting the Discount Rate: Ecocentrism and Environmental Economics." *Global Environmental Politics* 6, no. 4 (November 2006): 56–72.

Botcheva, Liliana. "Expertise and International Governance: Eastern Europe and Adoption of European Union Environmental Legislation." *Global Governance* 7, no. 2 (April–June 2001): 197–224.

Brown, Michael, Linda Treviño, and David Harrison. "Ethical Leadership: A Social Learning Perspective for Construct Development and Testing." *Organizational Behavior and Human Decision Processes* 97, no. 2 (July 2005): 117–134.

Burnell, Peter and Oliver Schlumberger. "Promoting Democracy – Promoting Autocracy? International Politics and National Political Regimes." *Contemporary Politics* 16, no. 1 (March 2010): 1–15.

Callioni, Patrick. *Compliance and Regulation in the International Financial Services Industry*. Middlesex: Global Professional Publishing, 2008.

Cardenas, Sonia. *Conflict and Compliance: State Responses to International Human Rights Pressure*. Philadelphia: University of Pennsylvania Press, 2010.

Carlin, Ryan and Ian Smith. "Regime Evolution: Democratization, Autocratization and Stability in the World 1981–2008." Presentation at the Annual Convention of the Southern Political Science Association, 2010.

Carrubba, Clifford. "Courts and Compliance in International Regulatory Regimes." *Journal of Politics* 67, no. 3 (August 2005): 669–689.

Chayes, Abram. "On Compliance." *International Organization* (Spring 1993): 175–205.
——. *The New Sovereignty: Compliance with International Regulatory Agreements*. Cambridge: Harvard Press, 1995.
Checkel, Jeffrey. "Norms, Institutions, and National Identity in Contemporary Europe." *International Studies Quarterly* 43 (1999): 83–144.
——. "Compliance and Conditionality." ARENA Working Paper no. 18 (2000).
——. "International Institutions and Socialization in the New Europe." ARENA Working Paper no. 11 (2001a).
——. "Taking Deliberation Seriously." ARENA Working Paper no. 14 (2001b).
——. "Why Comply: Social Learning and European Identity Change." *International Organization* 55, no. 3 (Summer 2001c): 553–588.
Checkel, Jeffrey and Andrew Moravcsik. "A Constructivist Research Program in EU Studies?" *European Union Politics* 2, no. 2 (June 2001): 213–231.
Cirtautas, Arista. "The Post-Leninist State: A Conceptual and Empirical Evaluation." *Communist and Post-communist Studies* 28, no. 4 (1995): 379–395.
Clarke, Ben. "Securing Compliance with International Humanitarian Law: The Promise and Limits of Contemporary Enforcement Mechanisms." *International Humanitarian Legal Studies* 1 (2010): 213–220.
Collinson, Simon and John Holden. "Decision-making and Market Orientation in the Internationalization Process of Small and Medium-sized Enterprises." *Management International Review* 45, no. 4 (2005): 413–436.
Cortell, Andrew and J. Davis. "How do International Institutions Matter? Domestic Impact of International Rules and Norms." *International Studies Quarterly* 40 (1996): 451–478.
——. "Understanding the Domestic Impact of International Norms." *International Studies Reviews* 2, no. 1 (Spring 2000): 65–87.
Cowhey, Peter. "Domestic Institutions and the Credibility of International Commitments." *International Organization* 47 (Spring 1993): 299–326.
Davies, Matthew. "Rhetorical Inaction? Compliance and the Human Rights Council of the United Nations." *Alternatives* 35 (2010): 449–468.
Dawisha, Karen and Bruce Parrott, eds. *Conflict, Cleavage, and Change in Central Asia and the Caucasus*. New York: Cambridge University Press, 1997a.
——. *The Consolidation of Democracy in East-central Europe*. New York: Cambridge University Press, 1997b.
——. *Democratic Changes and Authoritarian Reactions in Russia, Ukraine, Belarus, and Moldova*. New York: Cambridge University Press, 1997c.
——. *Politics, Power, and the Struggle for Democracy in South-east Europe*. New York: Cambridge University Press, 1997d.
Denk, Thomas and Derek Silander. "Problems in Paradise? Challenges to Future Democratization in Democratic States." *International Political Science Review* 33, no. 1 (January 2012): 25–40.
Diamond, Larry. *Developing Democracy: Toward Consolidation*. Baltimore: Johns Hopkins Press, 1999.
Dionne, Kim. "The Role of Executive Time Horizons in State Responses to AIDS in Africa." *Comparative Political Studies* 441, no. 1 (2011): 55–77.
Douglas, Mary and Aaron Wildavsky. *Risk and Culture: An Essay on the Selection of Technical and Environmental Dangers*. Berkeley: University of California Press, 1982
Downs, George, David Rocke, and Peter Barsoom. "Is the Good News about Compliance Good News about Cooperation?" *International Organization* 50, no. 3 (Summer 1996): 379–406.

Eckstein, Harry. *Regarding Politics: Essays on Political Theory, Stability and Change.* Berkeley: University of California Press, 1992.

Etner, J., M. Jeleva, and S. Spaeter. "Optimal Decisions on Energy Production with Heterogeneous Risk Perceptions." *French Council of Energy* (April 16, 2009).

European Bank of Reconstruction and Development, *Nuclear Safety Account General Agreement*, 1994.

Evans, Peter. *Embedded Autonomy*. Princeton: Princeton Press, 1995.

Fearon. James. "Bargaining, Enforcement, and International Cooperation." *International Organization* 52 (Spring 1998): 269–305.

Finn, John. "The Rule of Law and Judicial Independence in Newly Democratic Regimes." *Good Society* 13, no. 3 (2004): 12–16.

Finnemore, Martha. *National Interests in International Society*. New York: Cornell Press, 1996.

Finnemore, Martha and Kathryn Sikkink. "International Norm Dynamics and Political Change." *International Organization* 52, no. 4 (Autumn 1998): 891–893.

Foss, Nicole. "Nuclear Safety and International Governance: Russia and Eastern Europe." *Oxford Institute for Energy Studies* SP11 (June 1999): 1–100.

Geddes, Barbara. *Politicians Dilemma: Building State Capacity in Latin American*, Berkeley: University of California Press, 1994.

Gill, Graeme. *Democracy and Post-communism: Political Change in the Post-communist World*. New York: Routledge, 2002.

Goklany, Indur. *The Precautionary Principle: A Critical Appraisal*. Washington: CATO Institute, 2001.

Goldstein, Judith and Robert Keohane, eds. *Ideas and FP*. New York: Cornell Press, 1993.

Gorenburg, Dmitry. "The Rules of the Political Game in Russia." *Russian Politics and Law* 50, no. 3 (May–June 2012): 3–6.

Greenberg, Michael and Dona Schneider. "Gender Differences in Risk Perception: Effects Differ in Stressed vs. Non-stressed Environments." *Risk Analysis* 15, no. 4 (August 1995): 503–511.

Hale, Henry. "Formal Constitutions in Informal Politics: Institutions and Democratization in Post-Soviet Eurasia." *World Politics* 63, no. 4 (October 2011): 581–617.

Hall, John and John Ikenberry. *The State*. Minneapolis: University of Minnesota Press, 1989.

Helmke, Gretchen and Steven Levitsky. "Informal Institutions and Comparative Politics: A Research Agenda." *Perspectives on Politics* 2, no. 4 (December 2004): 725–740.

Herrmann, Richard and Vaughn Shannon. "Defending International Norms: The Role of Obligation, Material Interest, and Perception in Decision-making." *International Organization* 55, no. 3 (Summer 2001): 621–654.

Hertwig, Ralph, Thorsten Pachur, and Stephanie Kurzenhauser. "Judgments of Risk Frequencies: Tests of Possible Cognitive Mechanisms." *Journal of Experimental Psychology: Learning, Memory, and Cognition* 31, no. 4 (2005): 621–642

Hiskes, Richard. *Democracy, Risk, and Community: Technological Hazards and the Evolution of Liberalism*. New York: Oxford University Press, 1998.

Jaggers, Keith and Ted Robert Gurr. *Polity III: Regime Change and Political Authority; 1800–1994* (computer file). 2nd ICPSR version. Boulder, CO: Keith Jaggers, 1996

Johns, Fleur. "International Legal Theory: Snapshots from a Decade of International Legal Life." *Melbourne Journal of International Law* 2 (2009): 1–10.

Jönsson, Christer and Jonas Tallberg. "Compliance and Post-agreement Bargaining." *European Journal of International Relations* 4, no. 4 (1998): 371–408.

Katzenstein, Peter, ed. *The Culture of National Security*. New York: Columbia Press, 1996.

Keck, Margaret and Kathryn Sikkink. *Activists beyond Borders: Advocacy Networks in International Politics*. New York: Cornell Press, 1997.

Kirk, Elizabeth. "Noncompliance and the Development of Regimes Addressing Marine Pollution from Land-Based Activities." *Ocean Development & International Law* 39 (2008): 235–256.

Knox, John. "New Approach to Compliance with International Environmental Law." *Ecology Law Quarterly* 28, no. 1 (May 2001): 1–123

Koh, Harold. "The New Sovereignty: Compliance with International Regulatory Agreements." *Yale Law Journal* 106, no. 8 (July 1997): 2599–2659.

Kolben, Kevin. "The WTO Distraction." *Stanford Law and Policy Review* 21 (2010): 461–491.

Krebs, Ronald and Aaron Rapport. "International Relations and the Psychology of Time Horizons." *International Studies Quarterly* 56 (2012): 530–543.

LaPuente, Victor and Marina Nistotskaya. "To the Short-sighted Victor Belong the Spoils: Politics and Merit Adoption in Comparative Perspective." *Governance: An International Journal of Policy, Administration, and Institutions* 22, no. 3 (July 2009): 431–458.

Leeland, Matthew, ed. *International Policies and Issues of Nuclear Safety and Management*. New York: Nova Science Publishers, 2013.

Legro, Jeffrey. "Which Norms Matter? Revisiting the Failure of Internationalism." *International Organization* 51 (Winter 1997): 31–63.

Levi, Margaret. *Of Rule and Revenue*. Berkeley: University of California Press, 1988.

Linz, Juan and Alfred Stepan. *Problems of Democratic Transition and Consolidation*. Baltimore: Johns Hopkins Press, 1996.

Mann, Michael. "The Autonomous Power of the State: Its Origins, Mechanisms, and Result." *European Sociological Archive* 25 (1984): 185–213.

Marples, David. *Chernobyl and Nuclear Power in the USSR*. New York: St. Martin's Press, 1986.

McGillivray, Smith. "Trust and Cooperation through Agent-specific Punishments." *International Organization* 54, no. 4 (Autumn 2000): 809–824.

Medvedev, Grigori. *The Truth about Chernobyl*. New York: Basic Books, 1991.

Mitchell, Ronald. *Intentional Oil Pollution at Sea: Environmental Policy and Treaty Compliance*. Cambridge: MIT Press, 1988.

——. "Regime Design Matters: Intentional Oil Pollution and Treaty Compliance." *International Organization* 48, no. 3 (1994): 425–458.

Mitchell, Sara and Paul Hensel. "International Institutions and Compliance with Agreements." *American Journal of Political Science* 51, no. 4 (October 2007): 721–737.

Mushkat, Roda. Dissecting International Legal Compliance: An Unfinished Odyssey." *Denver Journal of International Law and Policy* 38, no. 1 (Winter 2009): 161–191

Nagle, John and Alison Mahr. *Democracy and Democratization: Post-communist Europe in Comparative Perspective*. London: SAGE Publications, 1999.

Naurin, Daniel and Anne Rasmussen. "New External Rules, New Internal Games: How the EU Institutions Respond when Inter-institutional Rules Change." *West European Politics* 34, no. 1 (January 2011): 1–7.

Nye, Joseph. "Nuclear Learning and US-Soviet Security Regimes." *International Organization* 41, no. 3 (Summer 1987): 371–402.

Oye, Ken. *Cooperation under Anarchy*. Princeton: Princeton University Press, 1986.

Paton, Douglas, Leigh Smith, and David Johnston. "Volcanic Hazards: Risk Perception and Preparedness." *New Zealand Journal of Psychology* 29, no. 2 (2000): 86–91.

Patton, Michael. *Qualitative Evaluation and Research Methods, Second Edition*. Newbury Park: Sage, 1990.
Payne, Rodger. "Persuasion, Frames, and Norm Construction." *European Journal of International Relations* 7, no. 1 (2001): 37–61.
Pickett, Susan. "Japan's Nuclear Energy Policy: From Firm Commitment to Difficult Dilemma Addressing Growing Stocks of Plutonium, Program Delays, Domestic Opposition and International Pressure." *Energy Policy* 30, no. 15 (2002): 1337–1355.
Powell, Emilia and Jeffrey Stanton. "Domestic Judicial Institutions and Human Rights Treaty Violation." *International Studies Quarterly* 53 (2009): 149–174.
Pridham, Geoffrey and Attila Agh. *Prospects for Democratic Consolidation in East-central Europe*. New York: Manchester University Press, 2001.
Princen, Thomas. "Long-term Decision-making: Biological and Psychological Evidence." *Global Environmental Politics* 9, no. 3, (August 2009): 9–19.
Przeworski, Adam. *Sustainable Democracy*. Cambridge: Cambridge Press, 1995.
Putnam, Robert. "Diplomacy and Domestic Politics: The Logic of Two-level Games." *International Organization* 42 (Summer 1988): 427–460.
Regan, Helen, Yakov Ben-Haim, Bill Langford, William Wilson, Per Lundberg, Sandy Andelman, and March Burgman. "Robust Decision-making under Severe Uncertainty for Conservation Management." *Ecological Applications* 15, no. 4 (2005): 1471–1477.
Risse, Thomas. "Let's Argue! Communicative Action in World Politics." *International Organization* 54, no. 1 (Winter 2000): 1–39.
Risse, Thomas, Stephen Ropp, and Kathryn Sikkink, eds. *The Power of Human Rights*. Cambridge: Cambridge Press, 1999.
Risse-Kappen, Thomas, ed. *Brining Transnationalism Back In: Non-State Actors, Domestic Structures, and International Institutions*. Cambridge: Cambridge Press, 1995.
Ruggie, John. "What Makes the World Hang Together? Neo-Utilitarianism and the Social Constructivist Challenge." *International Organization* 52, no. 4 (Autumn 1998): 855–885.
Sachs, Noah. "Beyond the Liability Wall: Strengthening Tort Remedies in International Environmental Law." *UCLA Law Review* 55 (2008): 837–904.
Schimmelfennig, Frank. "The Community Trap: Liberal Norms, Rhetorical Action and Eastern Enlargement of European Union." *International Organization* 55 (Winter 2001): 47–84.
Shelton, Dinah, ed. *Commitment and Compliance: The Role of Non-binding Norms in the International Legal System* Oxford: Oxford University Press, 2003.
Simmons, Beth. "Compliance with International Agreements" *Annual Review of Political Science* 1 (1998): 75–93.
——. "International Law and State Behavior: Commitment and Compliance in International Monetary Matters." *American Political Science Review* 94 (2000): 819–835.
——. "Capacity, Commitment, and Compliance: International Institutions and Territorial Disputes." International Relations Colloquium, University of Virginia, 2001.
Skerl, Jerca. "European Public Policy (With an Emphasis on Exequatur Proceedings)." *Journal of Private International Law* 7, no. 3 (2011): 461–490.
Skocpol, Theda, ed. *Brining the State Back In*. Cambridge: Cambridge Press, 1985.
Streich, Philip and Jack Levy. "Time Horizons, Discounting, and Intertemporal Choice." *Journal of Conflict Resolution* 51, no. 2 (April 2007): 199–226.
Sverdrup, Ulf. "Compliance and Conflict Management in the European Union: Nordic Exceptionalism." *Scandinavian Political Studies* 27, no. 1 (2004): 23–43.
Tejada, Matthew. *Bulgaria's Democratic Consolidation and the Kozloduy Nuclear Power Plant: The Unattainability of Closure*, edited by Andreas Umland. Stuttgart: ibidem-Verlag, 2005.

Thomas, Ward. "Norms and Security." *International Security* 25, no. 1 (Summer 2000): 105–139.

Underal, Arild. "Explaining Compliance and Defection: Three Models." *European Journal of Politics* 4 (March 1998): 5–30.

Vargo, George, ed. *The Chernobyl Accident: A Comprehensive Risk Assessment*. Columbus and Richland: Battelle Press, 2000.

Vezirgiannidou, Sevasti-Eleni. "The Climate Change Regime Post-Kyoto: Why Compliance is Important and How to Achieve It." *Global Environmental Politics* 9, no. 4 (November 2009): 41–63.

Victor, David, Kal Raustilia, and Eugene Skolnikoff, eds. *Implementation and Effectiveness of International Environmental Commitments*. Cambridge: MIT Press, 1998.

Viklund, Mattias. "Energy Policy Options – From the Perspective of Public Attitudes and Risk Perceptions." *Energy Policy* 32, no. 10 (2004): 1159–1171.

Weiss, Edith and Harold Jacobson. *Engaging Countries: Strengthening Compliance with International Environmental Accords*. Cambridge: MIT Press, 1998.

Wettestad, Jorgen. *Designing Effective Environmental Regimes: The Key Conditions*. Northampton: Edward Elgar Press, 1999.

——. "Designing Effective Environmental Regimes: Conditional Keys." *Global Governance* 7, no. 3 (July–September 2001): 317–341.

White, Brent. "Putting Aside the Rule of Law Myth: Corruption and the Case for Juries in Emerging Democracies." *Cornell International Law Journal* 43 (2010): 307–363.

Wilpert, Bernhard and Babette Fahlbruch. *System Safety: Challenges and Pitfalls of Intervention*. Bingley: Emerald Group Publishing, 2002.

Young, Oran. *Compliance and Public Authority*. Baltimore: Johns Hopkins Press, 1979.

Zartner, Dana and Jennifer Ramos. "Human Rights as Reputation Builder: Compliance with the Convention against Torture." *Human Rights Review* 12 (2011): 71–92.

Additional Reading

International Atomic Energy Agency. *RBMK Nuclear Power Plants Generic Safety Issues*. Vienna, Austria, October 1995.

——. *First Review Meeting of the Contracting Parties – Convention on Nuclear Safety*. Vienna, Austria, April 1999a.

——. *Report of the International Conference on Strengthening Nuclear Safety in Eastern Europe*. Vienna, Austria, June 1999b.

——. *Second Review Meeting of the Contracting Parties – Convention on Nuclear Safety*. Vienna, Austria, April 2001.

International Atomic Energy Agency General Conference. *Measures to Strengthen International Cooperation in Nuclear, Radiation and Waste Safety*. Vienna, Austria, August 2000.

Karatnycky, Adrian, Alexander Motyl, and Charles Graybow, eds. *Nations in Transit 1998: Civil Society, Democracy and Markets in East Central Europe and the Newly Independent States*. New Brunswick: Transaction Publications, 1999.

Karatnycky, Adrian, Alexander Motyl, and Aili Piano, eds. *Nations in Transit 1999–2000: Civil Society, Democracy and Markets in East Central Europe and the Newly Independent States*. New Brunswick: Transaction Publications, 2000.

Karatnycky, Adrian, Alexander Motyl, and Amanda Schnetzer, eds. *Nations in Transit 2001: Civil Society, Democracy and Markets in East Central Europe and the Newly Independent States*. New Brunswick: Transaction Publications, 2002.

——, eds. *Nations in Transit 2002: Civil Society, Democracy and Markets in East Central Europe and the Newly Independent States*. New Brunswick: Transaction Publications, 2003.

Karatnycky, Adrian, Alexander Motyl, and Boris Shor, eds. *Nations in Transit 1997: Civil Society, Democracy and Markets in East Central Europe and the Newly Independent States*. New Brunswick: Transaction Publications, 1998.

Nuclear Energy Institute. *Source Book: Soviet-designed Nuclear Power Plants in Russia, Ukraine, Lithuania, Armenia, the Czech Republic, the Slovak Republic, Hungary and Bulgaria, Fifth Edition*. Washington, DC, 1997.

Organization for Economic Cooperation and Development/Nuclear Energy Agency. "Chernobyl – Ten Years: Radiological and Health Impact." November 1995.

——. *Safety Research Needs for Russian-designed Reactors – Report by an OECD Support Group*. Paris, France, 1996.

——. *International Seminar in the Safety Research Needs for Russian-designed Reactors*. Tokyo, Japan, 1997.

——. *Annual Report*. Paris, France, 1998–2001.

——. *Overview of Nuclear Legislation in Central and Eastern Europe and in the NIS*. Paris, France, 1998.

——. *The Role of the Nuclear Regulator in Promoting and Evaluating Safety Culture*. Paris, France, June 1999a.

——. *Russian Minatom Nuclear Safety Research Strategic Plan – An International Review*. Paris, France, 1999b.

——. *Assuring Future Nuclear Safety Competencies*. Paris, France, 2000a.

——. *Regulatory Response: Strategies for Safety Culture Problems*. Paris, France, 2000b.

Stern, Linda, ed. *Freedom in the World 1999–2000*. Washington: Freedom House, 2000.

——. *Freedom in the World 2000–2001*. Washington: Freedom House, 2001a.

——. *Freedom in the World Country Ratings 1972–73 to 2000–01*. Washington: Freedom House 2001b.

——. *Freedom in the World 2001–2002*. Washington: Freedom House 2002.

Transparency International. *Corruption Perception Index, 1998–2002*.

——. *Global Corruption Report, 2001, 2003*.

United Nations. "Report to the United Nations Scientific Committee on the Effects of Atomic Radiation to the General Assembly, Annex J." 1999.

US Department of Energy, Office of International Nuclear Safety and Cooperation. *Monthly Activity Reports for Improving the Safety of Soviet-designed Nuclear Power Plants*. Washington, DC, 1996–2002.

——. *Measuring the Performance of US Efforts to Improve the Safety of Soviet-Designed Nuclear Power Plants: Annual Performance Measurement Reports for 1997, 1998, 1999, 2000, 2001*. Washington, DC, 1997–2001.

——. *Changes and Opportunities, Lessons Learned and Successes, Report Summarizing the January 1998, January 2000 Contractor Information Exchange for US Commercial Firms Involved in Improving Safety of Soviet-designed Nuclear Power Plants*. Washington, DC, 1998, 2000.

——. *Chernobyl – Improving Safety*. Washington, DC, January 1998a.

——. *Improving the Safety of Soviet-designed Nuclear Power Plants – Strategy Document – Rev. 1*. Washington, DC, June 1998b.

——. *Soviet-designed Nuclear Power Plant Profiles and Accomplishments*. Washington, DC, 1998–2001.

——. *Improving the Safety of Soviet-designed Nuclear Power Plants – Status Report. 1999.* Washington, DC, January 1999a.

——. *Presentation on Status of Safety Projects at Soviet-designed Nuclear Power Plants.* Washington, DC, August 1999b.

——. *Status Report – Ten Years of Safety Improvements. 2003.* Washington, DC, January 2003.

US Department of State, Bureau of Nonproliferation. "Nuclear Suppliers Group Fact Sheet." Accessed regularly 2002–2012. www.state.gov/t/np/rls/fs/index.cfm?/docid=3053.

US Nuclear Regulatory Commission Projects. Accessed regularly 2002–2012. www.state.gov/global/oes/science/science_agreement/nuclear_safety.htm.

2 Ukraine's uneasy past and uncertain future

"Chernobyl is the safest plant in Ukraine."

—Sergei Parashin, Chernobyl plant manager, reiterated by Oleksandr Moroz, Speaker of Parliament, December 1995

Introduction

In the early morning of April 26, 1986, fires broke out at the Chernobyl nuclear power plant in central Ukraine. The number-four reactor had been operating at low power levels when personnel lost control of the fission reaction and the reactor overloaded. Flames ignited the surrounding room, quickly spreading to the control room and throughout the building, where they easily overwhelmed the inadequate fire suppression systems. Operators shut down the other three reactors before the roof collapsed over unit 4, but unfortunately they could not prevent a radioactive cloud from spreading over the surrounding countryside. That radiation moved over Eastern and Northern Europe, eventually landing in Japan, Canada, and the United States. The horror of that day became even more apparent through the heroic efforts of Ukrainian fireman and construction workers who sacrificed their lives to plug the gaping hole.

Chernobyl became an ugly reminder of the inherent dangers of nuclear energy, and Soviet designed reactors in particular. Originally intended to operate for a minimum of thirty years, Western countries began pressuring the Ukrainians to close the facility as quickly as possible after the collapse of the Soviet Union, and in response to offers for Western political and economic assistance, the Ukrainian Supreme Rada (Parliament) declared that Chernobyl would close in 1993. However, as the deadline approached, the country's first President, Leonid Kravchuk, extended the time for another two years. The next President delayed even longer, declaring that Chernobyl would not close down permanently until the end of 2000. Western officials wondered in dismay why the Ukrainian government kept one of the most dangerous places in the world operating at all, let alone for more than a decade after the catastrophic disaster.

The original dates of 1993 and 1995 had been domestic commitments, and therefore not subject to any binding international resolutions. The 2000 timeline differed as it became the center of an extensive Western assistance program designed to improve safety throughout the country's aging nuclear industry, part of which required decommissioning Chernobyl before the end of its design lifetime sometime between 2007 and 2013. Accordingly, Ukrainian President Leonid Kuchma signed a Memorandum of Understanding (MoU) with the G7 and the European Commission in December 1995 establishing the conditions for closure in five years. One year later, a financing agreement with the European Bank for Reconstruction and Development (EBRD) confirmed the 2000 date.

Ukraine's initial commitments to close Chernobyl met with tremendous domestic resistance, yet the final international one faced the strongest opposition. Opposition from within the Ukrainian government, nuclear industry, and population created intense pressure to renege on the promise, and numerous officials argued all the way up to the millennium deadline that Chernobyl was vital to the economy and should not close down prematurely. Some even maintained that it should remain open indefinitely.

Kuchma too insisted that the West had failed in its part of the agreement by not adequately funding the construction of replacement power facilities or compensating Ukraine for billions of dollars in lost future revenue from Chernobyl's closure. He argued that Ukraine did not need to comply with its commitments either, pushing for continued operation at Chernobyl beyond 2000. However, his position found no basis in the international agreements he himself had signed, none of which established any Western prerequisites for Ukrainian compliance. The single clearest point stated that his government decommission Chernobyl by the end of 2000, which surprisingly it did ten days before the end of the year. This chapter addresses why the Ukrainians seemed to deny the very real possibility of another disaster occurring at Chernobyl, only to implement such an unpopular decision and close it on time.

Political instability and delayed compliance

The political system during Ukraine's first post-Soviet decade was tumultuous and unpredictable to say the least. Despite the passage of a new constitution designed to clarify the formal rules of the system, officials repeatedly attacked the structural delineation of power as Parliament and the President each tried to supplant the other's constitutional authority. Rules governing the election cycle were also widely contested and sometimes completely disregarded, adding even greater uncertainty to the system. In addition, informal rules of the game more often reflected clan rivalries than established bureaucratic practice. Yet even the clan-based rules remained unstable as groups fragmented and new officials acted as independent agents seeking the highest bidder rather than as part of an official political party. As a result, party discipline remained weak and day-to-day governance procedures unpredictable.

In addition, while widespread corruption within the system could have stabilized governance procedures as elites and citizens worked to establish alternative, non-state methods of getting things done, selective anti-corruption campaigns based on political rivalries added greater instability to the rules of the game as officials targeted their enemies rather than corruption itself. Worst of all for compliance with the country's international agreements, bureaucratic responsibility for the country's nuclear power plants changed frequently and unpredictably as part of the free-for-all that defined Ukraine in the 1990s. These conditions gave elites ample incentives to pursue short-term goals to protect current assets regardless of any promised future benefits of compliance. Under such conditions of rules instability, closing Chernobyl made no sense and the Ukrainians did everything possible not to do it.

Squandered hopes

Unlike other post-communist countries in Eastern Europe, Ukraine did not begin the constitutional process when it gained independence. Instead, the original Soviet constitution received additional wording to express the independence of the state and the creation of the executive branch, but little more reflected the presence of a new political order in Ukraine. Not surprisingly, the constitutional framework proved inadequate from the beginning as "even with dozens of amendments, it failed to define either the separation of powers between the legislature and the executive branches or their responsibilities."[1] However, the lack of clarity in the formal rules did not pose a serious problem for long-term thinking during the heady days of the independence movement in 1991 and early 1992 as unity within the dominant political clan enabled the government to make costly international commitments as a means of gaining Western support. In particular, the decision to sign the Non-Proliferation Treaty and give up a share of the Soviet nuclear arsenal in exchange for Western assistance led to greater Western offers for closing Chernobyl and improving the safety of Ukraine's other nuclear facilities.

The effective control of the political process by former Soviet elites (*nomenklatura*) during the 1991 independence referendum and the first presidential election may not have boded well for democratic government, but it offered a measure of stability in the transition away from an official one-party state. The beginnings of stabilization in the rules allowed a measure of risk taking and future thinking among the ruling elite, and this produced the initial decision to close Chernobyl in 1993. Unlike later political contests, the first presidential election produced a clear winner as former Communist Party ideology secretary Leonid Kravchuk easily defeated Vyacheslav Chornovil, leader of the Ukrainian nationalist party Rukh. Kravchuk's victory resulted in part from his unequivocal support for independence while maintaining close economic ties with Russia, but more importantly as a result of the financial and public support he received from the former state managers-turned-business owners who controlled Ukraine's economic future.

Kravchuk's patrons used their control over state assets to promote members of their group as political candidates, or when not possible, proxy representatives like the newly elected President. By doing so, they rapidly entrenched themselves across Ukraine's political, economic, and social landscape. Their success depended in large part on the lack of counter weights in the party system, namely the splintering of Rukh from an independence movement into a Ukrainian nationalist party that exacerbated ethnic divisions in the country. As a result, moderate nationalists competed with other democratic parties to form a de facto opposition to the Communists, neither of whom could balance the growing influence of the reconstituted *nomenklatura*. While the group maintained a loose affiliation with the Communist Party, self-interest made them more opportunistic than ideological. As a result, when popular discontent against the theft of state resources began to rise and threatened their position in power, party alignments began to change. Later, when opposition parties attempted to call for a referendum on new elections but fell short of the necessary three million-signature mark, the threat was sufficient for Kravchuk and his allies to react, and the President responded by removing Community Party leader Votid Fokin and installing the more moderate Leonid Kuchma as Prime Minister in October 1992.

However, Kuchma's selection did not represent a decline in the *nomenklatura*'s political influence per se, nor was his appointment a sign of more democratic forms of governance. He and Kravchuk each ruled with the support of powerful business elites from the eastern regions of Ukraine. Kravchuk's efforts to gain support from the Dnipropetrovsk clan by offering a key position to their man backfired though as the economy worsened and the electorate grew more impatient. Rather than forge a united front, *nomenklatura* clans began to turn on each other, adding instability to the informal rules of the game.

The ensuing unstable years

The political rift became obvious soon after Kuchma's appointment as his clan publically denounced the President and his supporters. Opposition parties quickly splintered in Parliament, and as a result of the mounting divisions within Parliament and between it and the government, "the majority of adopted laws were uncoordinated, random, and contradictory."[2] As political debate became more acrimonious, personalized, and unpredictable, elite time horizons shrank and immediate concerns trumped long-term planning. Unstable formal rules, identifiable through the lack of clear structural mechanisms for power transfer, the unclear delineation of delegated authority, and growing challenges to the delegations that existed at the time, created strong incentives for elites not to take political risks, whether in the form of compromise with opponents or through policies that would reduce their wealth and influence. Governance procedures also became confused due to the uncertainty as to who was in charge, how long they would be there, and what could be done to replace them, also shortening elite time horizons. As a result, "there were few moves towards political institutionalization in Ukraine"[3] and short-term thinking prevailed.

Short-term thinking affected Chernobyl directly as President Kravchuk's advisor for nuclear affairs and vice-chairman of the Academy of Sciences, Victor Baryaktar, backed out of the government's commitment to decommission Chernobyl in

1993, 1995, or anytime thereafter. He cited severe economic conditions, the high cost of fossil fuel supplies from Russia, high energy demand,[4] the instability of the national electricity grid, and a lack of domestic funds as reasons not to close Chernobyl. Some pro-Western politicians and business leaders tried publicly to assure international supporters that Ukraine remained committed to the 1993 deadline, but increasingly determined and widespread elite opposition soon eliminated that possibility. By April 1993, Deputy Director of the State Committee of Ukraine for Nuclear and Radiation Safety (GANU), Georgy Kopchinksy, had contravened Parliament's earlier degree for closure in 1993, stating that "no concrete decision [had] been made on the Chernobyl plant's future."[5]

GANU had been touted as an independent regulatory commission entrusted with ensuring the highest levels of operational safety at Ukraine's nuclear facilities. Instead, it served as a rubber stamp for the nuclear industry by simply endorsing plant management safety assessments since nuclear regulators had little de facto power to change conditions at Chernobyl even if they wanted to do so. Worse still, the agency became a pawn in the mounting clan rivalries, a fact Kopchinsky confirmed in September 1993 stating that "chances are growing that the Chernobyl nuclear power plant will continue operating beyond the end of this year. The country's political crisis is preventing any decisions on Ukrainian nuclear matters."[6]

President Kravchuk did very little to stabilize the situation. Several months earlier, he had staged a bloodless coup to concentrate all executive power in his hands. Prime Minister Kuchma and opposition parties in Parliament rebuffed the President's initial attack, but Kravchuk succeeded four months later with a presidential decree that effectively took control of the cabinet of ministers. The President had no constitutional authority to do so, and Kravchuk's enemies banded together across ideological lines in Parliament to assert their dominance at his expense, thereby reinforcing the lack of clarity in the constitutional separation of powers and a total disregard for the durability of the constitutional framework.

Kravchuk had also promised to call a referendum of confidence in himself and the government sometime that fall, yet when his mid-year approval ratings sank to below four percent, the President backed out by claiming that referenda were bad for the people, despite his reliance on the 1991 referendum to establish himself in the presidency. Referenda could have been a useful tool of government accountability, potentially stabilizing the rules of the game in the process. Unfortunately, they were used more often as a threat to shake rival elites than as a means to introduce predictable policy debate mechanisms.

To increase his popular appeal and divide his opponents, Kravchuk asked Parliament to reverse its earlier decision and allow Chernobyl to continue operating indefinitely. The President also called for restarting unit 2, even though it had been badly damaged by fire and could not operate without considerable repairs. The proposal passed the next day since growing popular discontent had everyone on edge and no elected official wanted to be seen endorsing a costly decommissioning plan that would have one immediate result political defeat for its supporters. While Ukrainian citizens had used the Chernobyl accident as a rallying cry for independence, the country's disastrous economic conditions presented a very real possibility of thousands freezing to death, and elites reasoned that lightning had already struck at Chernobyl and would not likely happen again anytime soon.

GANU Chairman Nikolai Steinberg supported the President's directive to keep Chernobyl running, but for reasons other than safety. His agency did not have the institutional authority to make autonomous decisions, and Steinberg owed his position to continued loyalty to the President. However, Steinberg surprisingly left the door open for early decommissioning on the condition that the West provide billions of dollars for construction of replacement power facilities, a completely unrealistically expectation at the time. Surprisingly and ultimately to the dismay of elites across the political divide, Western governments soon offered an enormous aid package to help the country increase nuclear safety at its remaining plants on the condition that Chernobyl close by the year 2000.

The severity of Chernobyl's problems should have been obvious, but the political system had become so uncertain that Ukrainian elites could not see the future dangers awaiting them when immediate ones dominated their perspective.[7] Atop the list of potential disasters sat the concrete shelter covering unit 4's ruined remains. Originally build to last until 2006, new evidence in 1993 showed the shelter would likely collapse ten years earlier without massive work to stabilize the roof and fill the nearly one thousand square meters of cracks over the surface of the sarcophagus.[8] Gaps had also opened at almost every joint connecting the shelter to the original building.[9] Worse still, a swirling cloud of radiation blew throughout the covered building, and radioactive isotopes leaked from cracks in the floor into the ground water below. Yet despite the obvious risks, Ukraine's nuclear regulators certified the existing structure as "acceptable to meet safety requirements."[10]

Around the same time, President Kravchuk signed the Convention on Nuclear Safety (CNS) requiring the creation of a sufficiently funded, autonomous nuclear regulatory authority. He had originally tasked GANU with overseeing nuclear safety in Ukraine, yet the agency lacked a predictable, reliable legal basis to give it formal autonomy, authority, or even the funds necessary to do its job.[11] As a result, regulators usually sided with their bureaucratic masters rather than offer objective safety assessments. One rare occasion of dissent had occurred the year before when GANU Deputy Director Kopchinsky openly criticized the government committee overseeing GANU as being concerned primarily with the production of electricity, not nuclear safety.[12] President Kravchuk responded by shutting regulators out of key decisions affecting Ukraine's nuclear facilities, and unilaterally deciding to rehabilitate unit 2 despite its considerable safety problems.

Persistent instability in the rules of the game shaped elite perspectives of all these problems such that long-term decisions became impossible, and policy implementation rarely had lasting effects. No single person bears the blame though. Even President Kravchuk was not alone in challenging the government's institutional framework. His one-time ally and later main rival, Leonid Kuchma also posed a serious threat to the stability of the political system, as did his own later rivals in Parliament.

Shortly after his appointment as Prime Minister in 1992, Kuchma declared that he could not do his job effectively without extraordinary powers granted by the legislature. Despite increasing criticism from President Kravchuk and the growing rift between their rival clans, the economic crisis had become so dire that Parliament gave Kuchma a free hand to fix it. Consequently, Kuchma's government effectively supplanted legislative functions with the right to issue sweeping decrees, even though the constitution forbade the use of such powers. The resulting economic collapse in 1993 quickly dashed whatever hopes might have rested on Kuchma's abilities to accomplish his stated goals, and the ruling Communist Party reasserted parliamentary authority, even taking away some of the Prime Minister's legal powers in the process. Early elections for Parliament and the presidency in 1994 could have resolved the crisis in governance, despite the questionable legality of both events under the existing constitution.

Given the rise in inter-clan rivalries and mounting opposition from new political groups before the elections, the *nomenklatura* used their clients in Parliament to pass a new electoral law that greatly reduced the chances of opposition victories. The law required candidates to win votes from at least fifty percent of voters registered with a party, not the more traditional fifty percent of votes cast, thereby eliminating independent voters from consideration. Most pro-democracy candidates failed to meet the threshold in the March 1994 elections, even in cities where they had the best chance of winning due to a Western-leaning electorate. The new requirement also delayed the appointment of several legislators so that "after some dozen runoff elections, the Kiev delegation to the Rada [remained] incomplete two years after the initial vote."[13] In the process, President Kravchuk had hoped low voter turnout and apathy would invalidate the first round of the presidential election. The people showed more initiative than he expected and Leonid Kuchma eventually defeated Leonid Kravchuk in the second round of voting. However, the *nomenklatura* kept control of the Parliament so that Kuchma's victory was not really a step towards greater democratization, rather simply a change in the President's last name. Worse still, the constitutional separation of power between the Parliament and the President remained unpredictable even though a new clan group had access to the President.

The political situation continued to destabilize, and in the face of deepening gridlock the next year, President Kuchma threatened to bypass the legislature and call a referendum to force the issue of a new constitution. Parliamentary Speaker Moroz responded by submitting a draft constitution that strengthened the legislature by stripping the President of the few defined powers he had under the current system. Kuchma retaliated with a proposal that gave the presidency tremendous power at the expense of the legislature.[14] Both sides wanted to reduce the authority of their rivals, illustrating not only the heightened animosity between government institutions, but also the lack of agreement on the fundamental division on power in the formal rules of the game.

In this midst of these political struggles, Ukrainian safety experts sounded the alarm on Chernobyl's weakening sarcophagus, confirming that it "could be destroyed by abundant snowfall, heavy rain, storms, or earthquakes . . . [with] a relatively high probability of tornadoes, which could aggravate the consequences of a collapse by lifting the radioactive dust to great height."[15] Shortly thereafter, a consortium of international safety experts and design manufacturers reported that in order to secure the shelter over unit 4, unit 3 also needed to shut down since both units shared key water piping, and the chemical treatment equipment for unit 3's cooling system rested underneath the plant's smoke stack – the most dangerous area of the sarcophagus. According to a consortium official, "if the stack collapsed, the same thing [would] happen to unit 3 as happened at unit 4" – catastrophic nuclear failure.[16] The European Commission and the EBRD responded immediately by proposing a massive Shelter Implementation Plan to fix the existing structure and build a new sarcophagus to encase both the damaged reactor and the failing shelter surrounding it. The program called for a Ukrainian commitment of nearly $200 million out of an estimated $800 million total bill.

Despite that commitment and the resulting efforts to improve nuclear safety in the country, work progressed slowly due to Ukrainian delays in implementing the Vienna Convention on nuclear liability,[17] without which Western governments would not allow their workers to engage in the most critical and dangerous safety work. The agreement protected foreign workers from lawsuits in the event of an accident, but the Ukrainian government said it could not present a liability bill to the Parliament for approval because legislators were occupied with trying to resolve the constitutional crisis between Parliament and the President.

To make matters worse, informal instability added to the unstable formal rules due to the uncertainty as to which government agency had authority over the nuclear industry, and unpredictability about how long that authority would last. This hindered Ukrainian compliance with the CNS provisions for establishing an autonomous regulatory agency as repeated bureaucratic restructuring under President Kuchma presented additional challenges to the already weak nuclear regulatory bureaucracies. Responsibilities changed without any predictable pattern, while redundant committees in Parliament struggled to wrest control of Ukraine's nuclear industry out of the government's hands as well.

The President had earlier dissolved GANU and the Ministry for Environmental Protection in December 1994 to create a new organization combining both groups as "a more effective system of regulating nuclear power and radiation safety, [but] supporters of an independent regulatory committee called the move a disaster."[18] Even after the change, the Chairman of the new Nuclear Regulatory Administration (ARA) still did not report to the President as stipulated in the CNS. Instead, the regulator relied on the Minister of Environmental Protection and Radiation and Nuclear Safety for political and budgetary patronage. Kuchma made the situation worse by introducing a new independent advisory committee in 1995 whose purpose was to counsel him on the operational costs and safety of Chernobyl, further diminishing the autonomy of ARA. He also delayed its funding and personnel approval for several years in violation of the later EBRD Grant Agreement.

In addition, Kuchma ordered the restructuring of the state utility to allow its merger with the owners of Ukraine's nuclear power plants. However, the new state utility Energoatom had no formal organizational structure within the government, nor did it receive authorization to do its job until February 1998. During that time, Kuchma reorganized the government infrastructure, including Energoatom and the ARA, two more times before 1998, and then once a year in 1999, 2000, and 2002. As a result, effective transitions were impossible and oversight became problematic at best, impossible at worst as nuclear facilities operated without licenses, ministry representatives asserted their control over regulatory matters, and plant management had special access to the cabinet that bypassed the ministry responsible for safety matters. Taken on their own none of these is totally damning, but as a whole they illustrate the high level of instability in the informal rules of the game, all of which enabled nuclear officials to circumvent what little government oversight existed in the nuclear industry while the President and his rivals in Parliament battled for supremacy.

Kuchma had a stronger position in those battles given his high approval ratings and Parliament's almost complete lack of legitimacy among voters, most of whom viewed it as the most corrupt and least trustworthy institution in Ukraine at the time.[19] Kuchma tried to break the deadlock by threatening to call for a referendum on early elections. In the face of a potentially disastrous showdown with the President, Parliament finally acquiesced to Kuchma's demand and pledged to confirm a final constitution sometime the next year. However, by mid-1996 Kuchma again challenged Parliament's authority by threatening to suspend it and impose emergency rule if the new constitution was not approved immediately.[20] Legislators gave in after an all-night session in June, even though they managed to impose a few restrictions on the President.

Kuchma appeared to have won, but the political battle threatened all sides because of the worsening economic conditions and the need for essential Western financial assistance to mollify a restless and angry electorate. The President signed an extensive international agreement that would provide those funds through the EBRD, and Parliament approved work on Chernobyl's new shelter as short elite time horizons viewed the possibility of total economic collapse as more risky then whatever political capital elites might gain by siding with the West on safety enhancements. In addition, while safety upgrades required a commitment to close Chernobyl, actual decommissioning could be delayed once the economy recovered in the short term, thereby giving elites political cover if the plan did not increase their popular support. Hence, when the first of two units closed in 1996, parliamentary officials, cabinet members, and plant management claimed it as a temporary safety refit.

Accordingly, Ukraine's systemic political instability hindered long-term thinking when considering costly decisions, and closing Chernobyl stood out as one of the most costly at the time. Despite some clarity regarding the formal rules themselves, their durability remained highly debated. In particular, Kuchma did not win total control of the system through his proposed "super-presidential" system because the new constitution de facto restricted his power to appoint and dismiss the government by requiring legislative approval to form a new government.[21] He could remove the government, but this would force a stalemate with rivals in Parliament that might move the electorate against him. Presidential decrees also possessed the force of law, but only in economic matters where no legislative statute already existed. Most importantly, Kuchma did not gain the right to dismiss Parliament arbitrarily. He could do so only if the legislature remained out of session for thirty days, which Kuchma tried to put into effect in July 1997 after Parliament recessed for the summer, a usage the provision most likely did not encompass. Midway through their break, the President announced he was considering dismissing Parliament and calling for early elections, most likely counting on the short turn-around time and limited news access of most legislators on vacation in order for his plan to succeed. In the end, it failed as Parliament members quickly "scurried back from Crimea to reconvene for a day on August 19 to deny the President the chance to attack the assembly."[22]

These problems resulted from more than personal antagonism or institutional rivalry; the constitution lacked an effective provision for resolving impasses between the President and Parliament, and even fewer reliable mechanisms for cooperation. Taking advantage of the lack of clarity in the formal rules, the President used his approval ratings to browbeat Parliament, and even Parliamentarians supportive of Kuchma's policies tended to vote against him out of self-defense. This situation could have stabilized the informal governance procedures if the protagonists allowed institutional identities to trump personal interest, enabling them to form predictable patterns, even in hostile relationships with other branches of government. Instead, each side regularly contested the boundaries of their authority, and given the ongoing challenges to the formal rules, elites quite reasonably had little confidence in the stability of the system. Surveys by the Rada's parliamentary Development Project in 1997 and 1998 showed that nearly three-quarters of the legislature believed the current crisis threatened the future of democratic governance, fifty-six percent believed the constitution could not provide a stable foundation for democracy, and nearly half blamed both the President and Parliament for the ongoing struggle.[23]

In addition to the lack of clearly delegated institutional authority and widespread disagreement about the durability of those delegations, power transfer mechanisms remained in flux as well. A new electoral law passed by the Rada in October 1997 changed the old proportional representation system by adding a requirement that fifty percent of parliamentary seats be chosen by a majority of voters. Kuchma opposed it because the new law would lead to a greater stabilization of the party system in Parliament, something that would invariably decrease the need for presidential decrees as a way of managing the economy.[24] He proposed extending the term of the existing Parliament and postponing elections, a power not granted to him by the constitution, and while his plan failed to stop the new law, it succeeded in continuing the instability in the rules of the game.[25]

Despite his prominent role, Kuchma does not bear sole responsibility for the chaotic situation at the time. His rivals in Parliament also bear the blame for Ukraine's problems as they initiated a motion to abolish the presidency after they won the 1998 Rada elections. The motion passed after three attempts and went to the Constitutional Court for review, which later overturned it.[26] In September that same year, Parliament took a more limited approach by voting to prohibit Kuchma from running for reelection. The decision had no legal power, but it succeeded in discrediting the President and weakening his popular support. As a result, the presidential elections in 1999 became a mudslinging competition to see who would be least dirty in the eyes of the electorate.[27] More importantly for the rules of the game, the election "put the question of the revision of the constitutional division of powers back on the agenda,"[28] as almost all of Kuchma's political opponents served as members of Parliament at the time.

Safety work at Chernobyl continued to suffer in the midst of these struggles. Ukrainian delays and excuses as to why they could not comply continued until Valentin Kupny of the Chernobyl Sarcophagus Project issued a report in 1999 analyzing the fire safety of the shelter. He stated categorically that the sarcophagus posed "the gravest fire hazard. The distribution of combustible material and the extent of their radioactive contamination are known only very approximately [projected at 2,000 tons] . . . [with] dangerously explosive concentrations of a number of components."[29] Perhaps the clarity of the report coming from a Ukrainian opened eyes to the truth because less than a week later, the first contract was signed to begin work on the sarcophagus later that year. Kupny's report also spurred Western donors to contribute the remaining money needed for the project, and while Ukraine's monetary contribution had been small to begin with, they could no longer blame Western governments for the delay. With incontrovertible evidence of the need to begin work on the shelter, the money came out of the Ukrainian budget.

Yet all sides of the political fight continued to cite the plant's vital role in building the economy even though energy demand fell throughout the 1990s due to low commercial production, with Chernobyl producing less than ten percent of the country's total generating capacity. Residential demand remained fairly constant, but the loss of numerous industries to idleness, decay, corporate fraud, and outright theft meant that Ukraine did not need as much power as it could produce even with Chernobyl operating at reduced levels.

Widespread non-payment for electricity also contributed to the unwillingness to close the plant on the belief that even small payments were better than none. This problem stemmed from the state utility Goskomatom, and its later incarnation Energoatom regularly accepting barter payments, with only about ten percent of all payments made in cash by 1999. The figure climbed to nearly fifty percent over the next few years, but customers still owed the state utility hundreds of millions of dollars in past-due bills as corruption within the state utility and plant management allowed partial payments and personal favors to keep the power flowing.[30]

As a result of insufficient revenue to fund full operations at the plant, workers' salaries started to lapse in the late 1990s, leading to protests and under-shifting for daily operations, thereby reducing regular maintenance schedules and slowing down the pace of safety modifications. In addition, Chernobyl plant managers sometimes temporarily shut down the facility or operated on partial power because they could not purchase fuel from Russia in time.[31] The 1986 disaster occurred while operating on partial power, and while normal operations at that level might not automatically cause another accident, the risks remained high given the plant's other serious problems. Chernobyl's unreliable power generation also added to the problems facing the country's electricity grid. Regular outages throughout the nuclear sector caused the grid to drop below the operational safety limit of 50 megahertz, threatening the entire system with collapse.

To compensate for these problems, the Ukrainians demanded Western payments for closing Chernobyl. The total decommissioning bill came to a staggering $4 billion, which they could not afford even if they had interest in doing so. Western donors responded with grants and social assistance programs, one of which built the International Chernobyl Center in Slavutych as a way to absorb plant workers after it closed, but assistance like this often came in the form of "in-kind" support rather than direct cash payments. Taking what the West offered, the Ukrainians persisted in demanding that replacement power should come from two nearly completed reactors at Khmelnitsky and Rovno, but during five years of EC and EBRD least-cost studies, the Ukrainians continually balked at the austerity measures required by the West to fund their completion. The EBRD plan called for an electricity rate hike of one quarter of one cent per kilowatt-hour to pay back the loan, while a comparable Russian plan required no price increase and cost about half as much. The Ukrainians opted to work with the Russians even though the savings came at the expense of vital safety modifications.

Even if the West had met the demand for full compensation, opposition to early closure would have remained high. Chernobyl plant manager Sergey Parashin remained one of the most ardent opponents of closing the plant anytime before 2007, a date closer to the original thirty year operational lifetime.[32] Political elites also resisted compliance as Kuchma's appointed Prime Minister, Yevgheny Marchuk, pressed for the plant to remain open. The President fired him two months later, but the next three Prime Ministers each called into question the need to close Chernobyl in 2000.

As the political system grew more chaotic, defying the terms of their international commitments gained popularity among embattled elites. Both sides of the political struggle publically criticized earlier decommissioning commitments, but instead of uniting against the West, they used their rivals' apparent "pro-Western" decisions to delegitimize them in the public eye. Kuchma succeeded more often than his enemies and continued to resist early decommissioning. Emboldened by a strong showing of popular support, the President kept up his public disagreements with the West, while fighting a more winnable contest against his domestic foes. To that end, he called for a national referendum in April 2000 to weaken Parliament. His proposal contained an expression of no-confidence in the current Parliament, followed by a measure to strip away parliamentary immunity to make selective prosecution more feasible. The third and fourth provisions went to the heart of the institutional struggle as Kuchma wanted to create an upper legislative chamber that the President would personally appoint, and reduce the size of the lower chamber from 450 to 300, making it more pliable to his will. A final proposal would give the President the right to disband Parliament if it failed to form a majority one month after the elections or approve the state budget within three months, a power that blatantly violated the existing constitution.

As with the earlier constitutional battles, the President's attacks could have formed a cohesive in-group identity among Parliamentarians, thereby introducing a measure of stability to the informal rules of the game. Instead, the party system remained weak and fragmented, and personal interests trumped institutional loyalty. Center-right legislators turned on the Communist Party Speaker of the Rada and eventually received enough support from independent legislators to force his resignation. In response, the Speaker obstinately refused to leave despite the legality of his ouster. "The parliamentary majority (243 deputies) walked out in protest and assembled in a separate session in another building. After two weeks of parallel existence, the majority (by that time 259) took over the Parliament building amidst skirmishes with leftist deputies."[33] The President took advantage of the surreal proceedings and his referendum passed.

Government bureaucracy also suffered from the tumultuous times. From 1997 to 2002, four different Prime Ministers governed Ukraine, and while transitions in power did not inherently destabilize the rules of the game, they did little to resolve the problems. Beginning with Kuchma's firing of Pavlo Lazarenko to stem his growing popularity and potential rivalry in the upcoming presidential election, Parliament then demanded the removal of his replacement, Valery Pustovoytenko six months later.[34] The President responded that if Parliament continued to oppose his government, he would dissolve the legislature and call for new elections, even though he had no legal power to do so. The anti-Kuchma bloc in the Rada countered on February 2 by approving impeachment proceedings, citing the President for daily violations of the constitution. A temporary de facto cease-fire lasted until the parliamentary elections in March 1998.

Pustovoytenko eventually succumbed to the pressure though, and resigned in December 1999 to head up the pro-Kuchma For Integral Ukraine Party. The President appointed another loyal clan member, Viktor Yushchenko as his replacement, but almost immediately his government came under attack by rival clans.[35] Divisions then began to form within Kuchma's own clan, and the President later withdrew his support for Yushchenko, who was forced out in 2001. Kuchma replaced him with yet another Dnipropetrovsk clan member, Antatoliy Kinakh, who survived the internecine battles for barely a year. In the midst of the continued fighting between political rivals, one persistent commonality was their resistance to comply with the requirement to close Chernobyl by the end of 2000.

Western officials countered their repeated refusals by insisting that closing Chernobyl as promised remained a necessary condition for further international cooperation. The message from Western Nuclear Safety Account donors, the European Commission, and the EBRD stayed the same as it had always been: close Chernobyl or risk losing future money. Ukraine needed IMF loans to stabilize the currency, EC and EBRD assistance to restructure the energy sector, and could not afford to fix all of Chernobyl's problems on its own. Yet elites repeatedly tried to call what they hoped would be the West's bluff. In the end however, the great debate ended with a whimper. While many would have run the risk of operating Chernobyl for years longer, the plant simply fell apart before it could be revived, and Kuchma finally signed its death warrant.

The end of one era and continuation of another . . .

Aside from running out of fuel in 2000, regulatory personnel discovered several hundred cracks in Chernobyl's coolant circuit and emergency cooling systems. Plant management wanted to keep the reactor running and fix the cracks along the way, even though the cracks were located in highly radioactive areas and workers would receive a collective dose of 6–10 Roentgen Equivalent Man (REM) in repairing a single breach (the annual limit for an individual was set at 5 REM). Management authorized piecemeal repairs, but inspectors discovered new cracks each time they reviewed the work. Eventually, so many cracks formed that the plant would have to be down for several months at a time to fix them, assuming funding became available. The West refused to help, and in the end, Chernobyl limped along until it finally ran so low on fuel that it had to operate on less than half-power just to make sure there was something to turn off at the December 21 closure ceremony.

Closure opponents fought until the bitter end to keep the plant running, despite the tremendous risks of operating an RBMK reactor at partial power. During a December 5 parliamentary hearing questioning the President's authority to close the plant, and further indicating the continuing uncertainty about institutional roles in the formal rules of the political system, legislators sought testimony from plant managers to justify continued operation at Chernobyl. Two French delegates attending the hearings reported they were "stunned and aghast by remarks of Chernobyl staffers who affirmed that their plant is safe and profitable and should continue to operate. [Those staffers] asked Ukrainian officials to strongly oppose the closure of Chernobyl 3, which they saw as imposed from abroad. It was as if nothing happened in 1986."[36] At the time, Ukraine's demands on the West remained largely unmet. Replacement facilities at the combined thermal plant in Slavutych had not been completed, and the government had received less than one-quarter of the nearly $3 billion in promised assistance from the West as international donors waited for Chernobyl to close before releasing the remaining funds to decommission it. Accordingly, Parliamentarians and Energoatom officials called early closure a costly mistake, and Kiev Mayor Alexander Omelchenko raised the possibility of modernizing and reopening the first and third reactors once the structure over the destroyed unit 4 had been secured.[37]

Yet despite these lingering problems, and after years of resisting his own commitments, President Kuchma publically condemned Chernobyl as a negative symbol that had no right to exist on earth. He stated at its closure ceremony that Ukraine willingly terminated Chernobyl following the highest priority interests of its people and the international community, adding that since Chernobyl was a threat to the entire world, Ukraine readily sacrificed part of its national interests for the sake of global safety.[38] However, regardless of his politically correct rhetoric, Kuchma did not terminate a healthy, viable nuclear power plant. Worn out and dilapidated, Chernobyl died of old age and the Ukrainians could do nothing to prevent it.

Even though Chernobyl closed on time, the Ukrainian case shows inconsistent low compliance. Elites regularly delayed safety improvements at Chernobyl, and at times disavowed their necessity altogether. Most of the country's remaining nuclear facilities underwent the necessary modifications, but the work lagged behind schedule for years due to Ukrainian unwillingness to comply with their obligations. Medium compliance also remained incomplete for years as the Ukrainians lacked an autonomous, sufficiently funded regulatory agency with the power to ensure safe operations throughout the country. President Kuchma's 2001 decree to create the Nuclear Regulatory Inspectorate (ARI) fulfilled part of the CNS requirement, and Regulatory Chairman Vadym Gryshchenko indicated that he had "the support of the President, Parliament, the Prime Minister and the cabinet. The President said that he [would] agree with all [our] decisions regarding all safety matters; safety comes first."[39] Unfortunately, many more years passed before regulators assumed their full responsibilities to confirm Gryshchenko's confidence. Finally, the Ukrainians failed to meet the conditions for high compliance because they did not establish and pay for a decommissioning fund as the necessary precursory step prior to shutting down Chernobyl, trying all available means to keep it running past the compliance deadline instead.

Low stability in the rules of the political game caused low compliance with Ukraine's international commitments, and while the latter condition became less relevant with Chernobyl's demise, instability in the rules persisted afterward as rival elites continued to battle for supremacy. Shortly after Chernobyl closed, parliamentary Democrats proposed a new electoral law based on a full proportional representation system designed to increase their chances in the otherwise rigged Ukrainian elections. Kuchma vetoed the bill five times until Parliament finally acquiesced to a mixed system. Six months later, the Tymoshenko bloc put forward a motion to impeach the President following the 2002 parliamentary elections. In the process, its proponents subverted constitutional provisions by usurping judicial authority over impeachment proceedings. The plotters charged that the President had been involved in attempted blackmail of collective farm managers to force their support. They also cited recordings of alleged telephone calls between Kuchma and members of the executive branch that implicated the President in the September 2000 kidnapping and murder of opposition journalist Georgiy Gongadze.

Government misconduct and corruption had never been limited to a few officials in Ukraine, instead functioning throughout the government and state bureaucracy so that aside from presidential and parliamentary rivalries, political elites also contended with the criminal-political nexus in Ukraine.[40] Patronage relationships between the legal and illegal sectors even led to the election of several convicted criminals to Parliament, with allegations of the same about ousted president Viktor Yanukovych. This widespread corruption not only demoralized society,[41] it also destabilized the informal rules of the game due in large part due to the selective enforcement of anti-corruption legislation. Elites charged and countercharged one another with corruption, regardless of the veracity of the claims. The case of former Prime Minister Yulia Tymoshenko stands out as the most blatant abuse of corruption laws going as far back as 2001 when President Kuchma initially supported her nomination in an attempt to divide the Hromada bloc in

Parliament, then when it became clear his plan had failed, fired her and had her arrested. President Yanukovych's efforts against her followed that pattern to the detriment of democratic development in the country.

Conclusion

It is quite clear that the Ukrainian political system experienced persistent instability during the period of compliance with international nuclear safety agreements. The formal and informal rules lacked clarity and predictability, exhibiting a) violence in the political arena, b) challenges to the delegation of constitutional authority, c) presidential threats of referenda to dissolve Parliament, d) counter threats to impeach the President, e) unreliable mechanisms of power transfer in the election cycle, f) frequent and unpredictable changes in bureaucratic responsibilities, g) criminal penetration of the multiple branches of government, and h) arbitrary use of anti-corruption measures to remove government ministers as part of elite clan rivalries. All of these made elite time horizons very short, and that short-term thinking butted up against compliance requirements to 1) improve nuclear safety at all the country's facilities, 2) develop and authorize an autonomous, capable nuclear regulatory agency to ensure those improvements occurred on time, and finally, 3) establish a decommissioning fund and close Chernobyl more than a decade before its original end-of-lifetime date.

Table 2.1 Ukraine in the 1990s

Rule Type	*Clarity*	*Predictability*
Formal	No	No
Informal	No	No

Conclusion: Rival elites in Parliament and the Executive branch challenged and sought to change the formal rules, both through direct constitutional revisions and indirect legal means. Responsibilities changed unpredictably. Informal rules operated arbitrarily with frequent challenges to patronage networks and ministerial appointments. Therefore, the system retained consistent **LOW STABILITY**.

Compliance

- Safety upgrades begun but not completed on time
- Nuclear regulatory agency not fully autonomous
- Closure based on inability to operate further, not due to decommissioning as a policy choice.

Conclusion: Despite the closure of the last reactor at Chernobyl, Ukraine did not meet the criteria necessary for high compliance. It achieved **LOW COMPLIANCE** through safety upgrades.

Instead of reaching full compliance by closing Chernobyl in 2000, instability in the formal and informal rules caused elites to opt for short-term goals at the expense of long-term benefits from cooperation with the West. Rules instability did not enable elites to see immediate costs as a necessary means to gain the much greater future benefits of compliance,[42] and as a result, compliance remained low throughout the decade. In the final analysis then, if the Ukrainians could have fixed Chernobyl's numerous problems, or even just the ones that prevented its continued operation, it is extremely likely the plant would still be running today, and that is a frightening thought.

Notes

1 Prizel (1997): 358.
2 Wolczuk (2001): 248–249.
3 Kubicek (2000): 47.
4 Western negotiating positions continually stressed falling energy demands as one reason Chernobyl need not operate any longer. They also stressed the need to increase energy efficiency by decentralizing heating systems so individual residences could set their own temperatures. The existing Soviet-era "one-size-fits-all" system prompted many open windows even during the coldest winter months.
5 *Nucleonics Week*, April 4, 1993: 13.
6 *Nucleonics Week*, September 16, 1993: 4.
7 Ukraine's nuclear power plants experienced a myriad of problems: a fire at Zaporizhya after a worker opened the wrong valve, extreme radiation overdoses at South Ukraine, Rovno management hiring workers with falsified work permits and education credentials, and blatant disregard of early warning systems (whenever they actually worked), not to mention the design flaws inherent in all Soviet-designed reactors.
8 *Nucleonics Week*, February 22, 1993: 15.
9 *Nucleonics Week*, February 25, 1993: 6.
10 *Nucleonics Week*, June 23, 1994: 16.
11 In 1994 Kopchinsky commented that "Ukraine still has no legal and regulatory basis for nuclear energy use – the Parliament either has no time to consider the bill GANU submitted over a year ago, or doesn't understand the importance of the measure. Not only does GANU not issue licenses to nuclear facilities – due to the lack of a licensing law – but safety audits of operating units are not completed and thus upgrade programs are on hold" (*Nucleonics Week*, April 7, 1994: 10).
12 *Nucleonics Week*, December 9, 1993: 11. Kopchinsky eventually resigned due to the increasing risks of continued operation. Citing the October 20 decision to allow open-ended operation for units one, two, and three he said, "I don't understand their decision. I spent long hours explaining the safety problems to government and Parliament officials. I could have accepted a decision on short-term operation, but not one essentially giving carte blanche to a plant considered unsafe not so long ago. I don't want to work in the safety committee under those conditions. The situation in Ukraine today is very similar to the situation in 1985–86 which led to the Chernobyl accident" (Ibid.).
13 Prizel (1997): 355.
14 "The most significant changes introduced in the Law on Power directly addressed the issue of control over the cabinet. These changes in the system were intended to terminate the Rada's powers of approval of the government as a whole and the power to affect the appointments of individual members of the cabinet." Wise and Pigenko, (1999): 43. Kuchma also sought to lessen Parliament's ability to vote no confidence in the government so that it had to wait a year after passing a provisional, limited no-confidence measure. However, while it "established a clearer control over the national government by introducing a disequilibrium in favor of the President . . . [it also] set the stage for the constitutional struggle, [as] it did not predetermine the outcome" (Ibid., 45–46).
15 *Nucleonics Week*, March 23, 1995: 1.
16 Ibid.

17 Schwartz (2006) reviews the development and implementation of nuclear safety liability and compensation conventions before and after the Chernobyl disaster. She accurately identifies the problem these laws have in attracting participants, an issue this book addresses by examining the costs and benefits of compliance.

18 *Nucleonics Week*, December 22, 1994: 8. Despite the administration's claims, Steinberg countered that "all the efforts that have been made over the past three years to establish a civilized nuclear safety regulatory regime in Ukraine may end up being wasted." Kopchinksy also called the President's decree "another example of an impulsive and authoritative approach to the nuclear power issue. The team of specialists that has been put together with hard labor is being ruined. Very few of them will want to be transferred to the new ministry" (*Nucleonics Week*, January 5, 1995: 12).

19 The 1997 Ministry of Justice and World Bank surveys of perceptions of service provision showed that "among a selection of government agencies, the Verhovna Rada, along with the Cabinet of Ministers, ranked negatively . . . tied for last place behind the local police and road traffic police. Once again, none of the government agencies rated positively, and few were ranked even near 'normal' on the scale" (Zyla, 1999: 256–257). See Government of Ukraine Ministry of Justice and World Bank Economic Development Institute (1997) for more details.

20 "It was also unclear whether Parliament alone could approve the constitution, or whether it needed approval through a popular referendum" (Kubicek, 2000: 57). Kuchma also tried to keep some of his "loyal" MPs from taking part in the debate so as to make his Law on Power the fall-back document when time ran out (Vorndran, 1999: 285).

21 See Ishiyama and Kennedy (2001): 1177–1191 for a solid treatment of the dangers of super-presidentialism in the former Soviet Union.

22 *East European Constitutional Review* (1997).

23 Wise and Pigenko (1999): 46–51.

24 Wolczuk (2001): 264.

25 The new law also failed to civilize the electoral campaign as physical violence increased in 1998 and 2002. Challengers assaulted incumbents – one Mayor took a beating in his own office – election rallies were marred by violence, and criminal elements gained seats in several local and national parliamentary bodies. See Council on Security and Cooperation in Europe (1998); *East European Constitutional Review* (1998); and Interparliamentary Union (2002) for more details.

26 The Constitutional Court showed itself to possess very little independence throughout this process and as a result, did not help to resolve the contest between the President and Parliament. Nor was the government able to escape the control of either institution, leaving it more often the lackey of either pole in much the same way as is the Court. See Wolczuk (2001): 263–264 for details on the debate concerning the Law on the Cabinet of Ministers and Kuchma's attempts to breach the constitution to gain more control over the government.

27 See Birch (2000): 106–107 for a breakdown of election results.

28 Wolczuk (2001): 265.

29 *Nucleonics Week*, February 25, 1999: 15.

30 These practices likely kept thousands from freezing, and with hyperinflation rampant at times, prospects for cash payments remained doubtful. However, corruption and waste added greatly to the overproduction of power and its subsequent non-payment.

31 The US Department of Energy proposed a plan in the late 1990s to assist Ukraine in developing nuclear fuel from domestic uranium deposits, but successive Ukrainian governments refused to change the country's reliance on Russia even though dependence made Ukraine vulnerable to shocks in the Russian economy, as happened in 1993 and 1998. That dependence also gave Moscow leverage over contentious issues, specifically the fate of the Black Sea Fleet and the rights of ethnic Russians in Crimea as pressure points in the 1990s, a preface to events in 2014.

32 *Nucleonics Week*, February 22, 1996: 15. A week later an official of the German Nuclear Safety and Environment Ministry commented that the Ukrainians "will never shut Chernobyl for another 15–20 years. Plant management is making plans to do serious backfitting with the help of Russian industry in a bid to keep the two operating units in service until at least 2010" (*Nucleonics Week*, February 29, 1996: 2).
33 Wolczuk (2001): 266.
34 State prosecutors charged former Prime Minister Lazarenko with illegally obtaining and laundering nearly $900 million and arranging three contract murders, including that of Parliamentarian Yevgen Scherban. He did not deny the charges, but rather implicated other top political officials, including the presidential administration, which he claimed had embezzled millions from IMF loans.
35 He later became the leader of the 2004 Orange Revolution, only later to lose his reelection bid in disgrace for failing to follow through on the Revolution's central promises to end corruption, increase government accountability, and improve the lives of Ukraine's economically and politically marginalized masses. He faced a tall order, but fighting with his original supporters in Parliament enabled the opposition to stymie whatever good he might otherwise have accomplished.
36 *Nucleonics Week*, December 14, 2000: 13.
37 *Nucleonics Week*, December 21, 2000: 12.
38 *Nucleonics Week*, December 21, 2000: 11–12.
39 Vadym Gryshchenko, in discussion with the author, September 2002.
40 Shelly (1998): 651. In addition, Transparency International ranked Ukraine eighty-seventh out of ninety-one in 2000, and eighty-third in 2001.
41 "Only 7% of the Ukrainians believe that the government is doing a good job in fighting organized crime, whereas 87% believe the contrary" (Ibid., 649).
42 The same can be said of the government's decision to accept Russian financial backing at the expense of EU support.

References

Birch, Sarah. *Elections and Democratization in Ukraine*. New York: Macmillan Press, 2000.

International Atomic Energy Agency (IAEA). Convention on Nuclear Safety, 1994.

Council on Security and Cooperation in Europe. "Report on the March 29, 1998 Parliamentary Elections."

East European Constitutional Review. "Constitution Watch: Ukraine." 6, no. 4 (Fall 1997); 11, no. 1 and 2, (Winter and Spring, 1998).

Government of Ukraine Ministry of Justice and World Bank Economic Development Institute. *Ukraine National Integrity Survey: Citizen's Experiences of Public Service Quality, Integrity and Corruption*. 1997.

Inter-parliamentary Union. "Ukraine Last Elections, March 31, 2002." Accessed regularly 2002–2012. www.ipu.org/parline-e/reports/2331_E.htm.

Ishiyama, John and Ryan Kennedy. "Superpresidentialism and Political Party Development in Russia, Ukraine, Armenia, and Kyrgyzstan." *Europe-Asia Studies* 53, no. 8 (2001): 1177–1191.

Kubicek, Paul. *Unbroken Ties: The State, Interest Associations, and Corporatism in Post-Soviet Ukraine*. Ann Arbor: University of Michigan Press, 2000.

Nucleonics Week, February 22, 1993: 15; February 25, 1993: 6; April 4, 1993: 13; September 16, 1993: 4; December 9, 1993: 11; April 7, 1994: 10; June 23, 1994: 16; December 22, 1994: 8; January 5, 1995: 12; March 23, 1995: 1; February 22, 1996: 15; February 29, 1996: 2; February 25, 1999: 15; December 14, 2000: 13; December 21, 2000: 11–12.

Prizel, Ilya. "Ukraine between Proto-democracy and 'Soft' Authoritarianism." In *Democratic Changes and Authoritarian Reactions in Russia, Ukraine, Belarus, and Moldova*, edited by Karen Dawisha and Bruce Parrott, chapter 8. New York: Cambridge University Press, 1997.

Schwartz, Julia. "International Nuclear Third Party Liability Law: The Response to Chernobyl." *International Nuclear Law in the Post-Chernobyl Period – OECD* (2006): 37–72.

Shelly, Louise. "Organized Crime and Corruption in Ukraine: Impediments to the Development of a Free Market Economy." *Demokratizatsiya* 6, no. 4 (Fall 1998): 648–663.

Vorndran, Oliver. "Institutional Power and Ideology in the Ukrainian Constitutional Process." In *State and Institution Building in Ukraine*, edited by Taras Kuzios, Robert Kravchuk, and Paul D'Anieri, chapter 10. New York: St. Martin's Press, 1999.

Wise, Charles and Volodymyr Pigenko. "The Separation of Powers Puzzle in Ukraine: Sorting Out Responsibilities and Relationships between President, Parliament and the Prime Minister." In *State and Institution Building in Ukraine*, edited by Taras Kuzios, Robert Kravchuk, and Paul D'Anieri, chapter 2. New York: St. Martin's Press, 1999.

Wolczuk, Kataryna. "Ukraine: Tormented Constitution-making." In *Democratic Consolidation in Eastern Europe: International and Transnational Factors, Volume 2*, edited by Jan Zielonka and Alex Pravda, chapter 9. New York: Oxford University Press, 2001.

Zyla, Roman. "Corruption in Ukraine: Between Perceptions and Realities." In *State and Institution Building in Ukraine*, edited by Taras Kuzios, Robert Kravchuk, and Paul D'Anieri, chapter 9. New York: St. Martin's Press, 1999.

Additional Reading

Commission on Security and Cooperation in Europe. "Ukraine's Parliamentary Elections, March 29, 1998." (April 1998).

——. "Ukraine's Presidential Elections, October 31 and November 14, 1999." (December 1999).

D'Anieri, Paul. "The Impact of Domestic Divisions on Ukrainian Foreign Policy: Ukraine as a 'Weak State'." In *State and Institution Building in Ukraine*, edited by Taras Kuzio, Robert Kravchuk, and Paul D'Anieri, chapter 4. New York: St. Martin's Press, 1999.

Diuk, Nadia. "Sovereignty and Uncertainty in Ukraine." *Journal of Democracy* 12, no. 4 (October 2001): 57–65.

Garnett, Sherman. "Like Oil and Water: Ukraine's External Westernization and Internal Stagnation." In *State and Institution Building in Ukraine*, edited by Taras Kuzio, Robert Kravchuk, and Paul D'Anieri, chapter 5. New York: St. Martin's Press, 1999.

Grant Agreement between the Government of Ukraine and the European Bank for Reconstruction and Development, Chernobyl Nuclear Safety Project, 1996.

Kuzio, Taras. "Belarus and Ukraine: Democracy Building in a Grey Security Zone." In *Democratic Consolidation in Eastern Europe: International and Transnational Factors, Volume 2*, edited by Jan Zielonka and Alex Pravda, chapter 17. New York: Oxford University Press, 2001.

Memorandum of Understanding between the Governments of Ukraine, the G7 and the European Commission, December 20, 1995.

Sushko, Olexsandr. "The 2002 Parliamentary Elections as an Indicator of the Sociopolitical Development of Ukraine." *Demokratizatsiya* 10, no. 4 (Fall 2002): 568–577.

Wise, Charles and Trevor Brown. "The Consolidation of Democracy in Ukraine." *Democratization* 5, no. 1 (Spring 1998): 116–137

3 Armenia at the crossroads

"We can tolerate the risk of running Metsamor because other risks are intolerable, they are much higher."

—Jivan Tabibyan, Armenian Ambassador and Chief of Mission to the Organization for Security and Cooperation in Europe, September 2002

Introduction

The nuclear power plant at Metsamor sits on an active seismic fault, the site of a devastating earthquake that killed more than fifty thousand people in 1988. Widespread popular demands forced the plant to close even though it was not seriously damaged because greater details about the Chernobyl disaster had been released, and fears abounded it could happen in Armenia. However, the plant would not stay down for long as plans began in 1993 to restart the facility to meet the country's dire economic needs. The country faced catastrophic economic collapse and immediate problems began to outweigh the potential risks of a nuclear disaster.

Since gaining independence in 1991, Armenia struggled to meet its most basic energy needs. Conflict with Azerbaijan, a trade embargo by Turkey, and irregular fuel shipments through Georgia denied Armenia any reliable alternative to Metsamor. Making matters worse, during the first two years of independence, the country experienced massive unemployment, starvation, and unprecedented emigration to the West and Russia. "It was as close to hell as we want to get," according to Ambassador Tabibyan.[1] Equally problematic, over-reliance on the Lake Sevan hydroelectric plant caused widespread ecological damage. Water levels fell over eighteen feet in two years, turning the once beautiful lake into a swap and seriously threatening the capital city's water supply. The efforts produced a meager one hour of electricity per day in Yerevan. Furthering the long-term environmental damage, Armenians also chose to chop down their ancestral woodlands to keep from freezing in the winter, increasing soil erosion and flooding in the springtime.

One man rose to dominate those dark days. A war veteran with strong military ties and nationalist appeals, President Levon Ter-Petrossyan strengthened his grip on the political system as the country slid into economic and social chaos. However, the President owed much of his power to continued support from the military and economic ministries, and as a result, over time their patronage relationship prevented the worst of arbitrary rule and began to stabilize the rules governing post-independence Armenia. In particular, the political system began to develop clear methods of power transfer for the President and Parliament through increasingly predictable election laws, and a new constitution in 1995 further entrenched the preeminence of presidential power.

In the context of stabilizing political rules, the government of Armenia signed a Partnership and Cooperation Agreement (PCA) with the European Commission promising to close Metsamor by 2004. In return, the West offered increased trade, monetary assistance, and infrastructural development if Armenia also complied with short-term safety improvements and the development of an autonomous nuclear regulatory agency. Unfortunately, the plant was in extremely poor condition at the time as safety had deteriorated since its initial closure in 1989, and parts had to be cannibalized between the units to keep one of them running. The Armenians could not afford to do everything on their own so Western assistance allocated $500 million toward necessary upgrades. However, the timeline for those upgrades added several years to Metsamor's anticipated restart, postponing vital energy production for the country. Immediate aid tried to fill the gap by preparing alternative energy sources, nearly $57 million going to upgrade a 300-megawatt fossil fuel plant and transmission line to the capital, but these took time to develop and would not alleviate growing worries about popular unrest under the harsh winters and rampant economic decline. How then could the Armenian government make such a costly commitment to delay restarting Metsamor at that time?

Despite being caught between the irresistible force of impending social disaster and the immovable object of Western conditions for assistance, Armenian political elites began to prioritize long-term benefits of cooperation over short-term economic needs. They could have simply ordered the restart regardless of safety concerns using whatever money was available; the need was that great since Metsamor accounted for nearly half of the country's energy production capacity at the time. However, the political rules were progressively stabilizing so elites could look favorably on costly short-term sacrifices to gain the greater long-term rewards of Western assistance.

This continued for several years as both rules stability and compliance moved higher due to the President and his clique stabilizing the political system and putting in place the initial safety improvements at Metsamor. Unfortunately, just as those improvements began to make a marked difference, the political system started a precipitous decline toward instability and short-term thinking came to prevail. The seminal event occurred in February 1998 when President Ter-Petrossyan resigned under pressure from his former political ally, Defense Minister Vazgen Sargsyan. The once clear hierarchy of power had been replaced, and the ensuing period of political instability culminated in Prime Minister Sargsyan's murder one year later.

The prime minister's murder exacerbated the early signs of competition between branches of government and different elite patronage networks within them. The conflict appeared to end in May 2000 with the resurgence of the presidency as the strongest government institution and President Robert Kocharyan as the chief political patron, yet challenges continued for several years, producing a measure of rules stability but not enough to enable high compliance regarding the country's international nuclear safety agreements.

Western governments acknowledged that reality, and since the PCA lacked enforcement mechanisms in general, they amended the terms to allow for a 2006–2007 closing date. Yet even those dates came and went as the government decreed in 2012 and 2013 that the facility would remain open past 2020.[2] As a result of the revisions to the PCA and the decision to extend Metsamor's lifetime of operation, Armenian elites have only been able to achieve medium compliance, a feat they successfully reached in the last decade.

Stability and compliance: A tale of two Armenias

September 1991 offered the promise of hope for Armenia as the vast majority of citizens supported a peaceful transition out of the Soviet Union and toward democratic government thereafter. The ruling Communists responded by calling for a public referendum, and the Armenian National Movement (HHS) headed by war hero Levon Ter-Petrossyan emerged as the dominant political party. HHS's victory in the previous year's election for the Supreme Soviet paved the way for Ter-Petrossyan to win the first presidential election one month after the referendum. The new office limited the authority of the carry-over Parliament, but there were no serious challenges from the legislature as elites accepted the new institutional division of power as beneficial for the country.[3] As a result, the formal rules remained largely uncontested for the first four years of independence, and the informal rules became more stable through the President's patronage of the defense and economy ministries. The rules were not completely inviolable since the country was in the midst of a transition from communist rule, but the direction of the change appeared fairly predictable.[4]

Despite the promising developments in the political rules of the game though, the country entered a period of devastating economic and social crisis. The Georgian civil war all but eliminated regular fuel shipments from abroad and upset the fragile stability of the Caucasus region. Nagorno-Karabakh erupted in armed conflict as Armenians drove local Azerbaijanis out of the area, to be followed by forced relocation from both countries shortly thereafter. Turkey responded by imposing an embargo on all goods and services to Armenia, effectively cutting off Western trade with the country. Russia and Iran remained the only two neighboring countries positioned to assist Armenia, but both experienced their own economic difficulties and assistance remained limited and sporadic as a result.

Cut off from alternative energy sources, Armenia began running the Lake Sevan hydroelectric plant beyond capacity, dropping the water level precipitously and causing serious health concerns in Yerevan. The sacrifice of the largest fresh water source in Armenia barely made a dent in the country's energy needs. Meanwhile, hundreds of megawatts of generating capacity lay dormant as the country slipped further into the abyss.

The decision to restart Metsamor was not an easy one because of fears of another Chernobyl. Those fears made sense given the poor safety conditions at the plant as it suffered from the same deficiencies present in all early Soviet-designed reactors: poor management, insufficient fire control, outdated computers, lax training, bad construction, and no containment structure to prevent the catastrophic release of radiation in the likely event of an accident. As a VVER 440/230 design, the plant had greater safety than the RBMK models, and it had received some modifications to protect against a high-level seismic event. However, Metsamor still had serious problems to overcome in order to prevent a disaster.

Politicians and nuclear officials debated for nearly two years until they reached a formal decision to restart the reactor in April 1993. Domestic conditions were terrible, but despite their reservations, the likelihood of an accident occurring seemed less important than the more tangible needs facing the country. In what turned out to be a bit of revisionist history, the same politicians who had earlier bowed to public pressure and argued that the plant was unsafe in 1989, retreated and labeled "the 1989 decision hasty and wrong, taken in an emotional context."[5] Western safety experts countered that restarting Metsamor would be reckless and foolhardy, but compliance did not enter the discussion since Armenia had not yet signed any international agreements regarding Metsamor.

One man, one rule

Despite myriad problems facing the country during the early 1990s, stability in the rules of the game was not a primary concern. The introduction of the presidency in late 1991 had worked well as post-communist elites accommodated to the new institution as part of their newly gained independence from the Soviet Union. Strong political families, regional affiliations, and clan-based identities added to the positive side of the political ledger.[6] Patronage continued to define the allocation of political positions and economic resources as it had for generations before, enabling elites to count on familiar methods of determining who governed what, how they did it, and for how long. The presence of clientelism hindered the development of liberal democracy, but it helped stabilize the informal rules of the political system and the Armenians needed all the good news they could get at the time.

The formal rules were also reasonably stable under Ter-Petrossyan even though he did not have complete control of the political system. With an election mandate of over eighty percent in 1991, the political costs of trying to challenge him or the rules that favored his position remained high. However, those calculations began to change in 1995 with plans to increase his power over the legislature through a constitutional referendum. In response, a few opposition groups offered marginal criticism and presented an alternative draft to the electorate, but the President's position as top political patron enabled Ter-Petrossyan to squash them easily. He overruled the opposition through his clients in the Constitutional Court, which summarily denied their petition, and took his case directly to the people. The President also publically chastised the opposition as a party of war bent on dragging the country into an even greater conflict with Azerbaijan and Turkey, and accusing them of trying to destabilize the government, something his new constitution would prevent.

Ter-Petrossyan exploited the country's weak civil society, and his party and the new constitution won by a sizeable majority.[7] As a result, the President used his control over government appointments to push through painful economic measures advocated by the West, despite their disastrous effect on household incomes and living standards.[8] Ironically, as Ter-Petrossyan attempted to bring the economy more in line with the West, he became more of an international pariah because of his increasingly autocratic control over the political system. The President had drawn sharp criticism from the West when he ordered the suspension of the Armenian Revolutionary Federation (Dashnaktsutyun) before the 1995 elections to minimize their influence on the constitutional debate. His heavy-handedness turned some voters against him, and the seeds of a credibility gap began to grow for the long-time populist. In the meantime, Ter-Petrossyan prevented an alternative power base from developing in the government by regularly changing Prime Ministers in an attempt to entrench his control over the patronage system in Parliament.[9]

The new constitution passed, and while it placed some limits on the President's ability to act arbitrarily and contrarily to the constitution, Ter-Petrossyan remained largely unconstrained. Most importantly, the new system complemented the existing division of power between Parliament and the President since a measure of continuity ran through the process – the President had dominated the political system since 1991 and the new constitution entrenched existing de facto powers rather than created new ones ex nihilo. In particular, Article 55 enabled the President to dissolve the National Assembly; appoint and remove the Prime Minister, his cabinet, all judges from the Constitutional Court to local courts, and the prosecutor general; and prevent new parliamentary elections in times of national emergency. The President also effectively controlled the budget through his appointees to the government.[10]

The new constitution protected some legislative rights but the balance clearly favored the executive. Articles 62 and 63 gave the National Assembly priority in law making, but Article 56 countered that power by allowing for binding presidential decrees. Even so, MPs could overturn a presidential veto with a simple majority, and votes of no-confidence would force his appointed Prime Minister to resign. Parliament could also force the President to resign, but only after he had been convicted by the Constitutional Court of treason or other high crimes against the state. This seemed unlikely as long as the President controlled appointment and dismissal to the Court and had the support of the underlying power ministries in Parliament.

The constitutional mechanisms for power transfer also clarified how and when parliamentary elections could be held, specifically by ruling out early elections except after dissolution of Parliament by the President and by granting immunity to sitting legislators. These measures instilled a sense of predictability to the rules of the game, and diminished the threat of politically motivated criminal charges.

While immunity ceased if the President dissolved the legislature, the overall trajectory of the system moved toward greater stability with Ter-Petrossyan at the helm.

Accordingly, Ter-Petrossyan appointed relatively compliant Prime Ministers to support his economic reforms and firmly set Armenia on the path of economic liberalization. Even though widespread corruption overshadowed the privatization of state enterprises, the informal rules of the game operated smoothly as patrons rewarded loyal clients with political positions and economic resources. Unfortunately, despite the positive stabilization of the rules, the economy continued to struggle with a dramatic impact on Metsamor. Limited government resources meant the facility's safety measures could not be tested adequately prior to restart, increasing the risks of disaster even as the country became more desperate to use it.

Metsamor's untimely revival

International safety experts had begun to worry about growing domestic pressure to restart Metsamor shortly after independence from the Soviet Union. Alexander Kalugin of the Russian Kurchatov Institute (one of the principal Soviet reactor design institutes) confirmed that "insufficient safety levels [had originally] signed the death warrant of Metsamor 1 and 2."[11] Western safety experts went further to say that safety had actually degraded since then; an IAEA letter to the Armenian Energy Minister cited major problems and called the situation categorically "not good."[12] When asked if he thought the plant should remain closed and if the Armenians would listen to Western advice, World Association of Nuclear Operators President Walter Marshall said the plant would come back on line "no matter what shape it [was] in."[13]

With domestic needs mounting every day, political elites went ahead with plans to restart Metsamor, and the formal announcement came in 1993 to restart unit 2 after stripping its counterpart for spare parts. The West offered to help but conditioned assistance on nearly $500 million in safety upgrades. The implementation of the extensive program invariably delayed Metsamor's restart, but rules stability played an essential role in elite decision making at that critical juncture. Even though political elites were caught between impending social disaster and strict conditions for assistance, they pursued a longer-term perspective regarding safety than their immediate needs dictated. Officials could have ordered the restart regardless of safety concerns using whatever means at their disposal, as would be expected if they had the risk acceptant safety culture Western officials feared.[14] However, safety culture had not become more risk acceptant, as evidence by the lengthy debates and acceptance of even further delays due to safety upgrades. Instead, elite time horizons had lengthened under the stabilizing political rules, thereby making it possible to endure some short-term costs so as to receive the longer-term benefits of cooperation with the West. Agreeing to Western demands did not mean the Armenians were completely passive about the arrangements though. Government officials regularly tried to squeeze more money out of international donors while waiting to restart Metsamor, even while accepting the requirements to comply. Compliance offered the only way to build long-term relationships with the West, and the Armenians needed those connections to ensure the country's development for the future.

Primary work focused on badly needed safety upgrades, most of which were fundamentally necessary for restart. Western officials also argued for long-term replacement sources for Metsamor, and the European Bank for Reconstruction and Development (EBRD) conditioned further funding on delaying Metsamor's restart until completion of all vital safety upgrades, hoping "to avoid the need to restart Metsamor. The agreement allow[ed] the bank to back out and demand repayment if the plant restart[ed] and [was] deemed unsafe."[15]

The money began to flow after Armenia signed the Convention on Nuclear Safety (CNS) in 1994 and President Ter-Petrossyan created the Armenian Nuclear Regulatory Authority (ANRA) to ensure compliance with CNS safety requirements. However, ANRA struggled to ensure safe operation with limited funds from the central government and insufficient training for personnel, thereby calling into question the country's overall compliance behavior.[16] Assistance from international donors made up the shortfall, but the CNS clearly laid the responsibility for maintaining sufficient regulatory capacity on the government. Therefore, ARNA was a step in the right direction but needed more support to be truly effective.

Improved safety and wishful thinking could not keep Metsamor from starting under a cloud of doubt in October 1995 though. Questions lingered about the safety of the plant's pressure vessels and seismic upgrades, and Western officials remained skeptical about its long-term viability. Instead, they wanted it shut down permanently. The EBRD and the European Union began moving in that direction by conditioning all future assistance on early closure in 2004.[17] That date became firmly established in the 1996 Partnership and Cooperation Agreement (PCA) between the European Commission and Armenia, although the Armenians included a clause conditioning shutdown on the presence of replacement power sources.[18] In response, the EBRD funded the Hrazdan fossil fuel plant to meet that condition, and the Bank pressured the Armenians to increase electricity prices to supplement the cost. International donors also worked to settle the country's disputes with Azerbaijan and Turkey to free up oil and gas shipments. These measures were designed to reduce Armenia's dependence on Metsamor and lock the government into decommissioning the plant by 2004.

The abrupt decline of rules stability

Western support did not carry over to every interaction with Armenia though, as criticism of Ter-Petrossyan's regime grew louder after allegations of election fraud increased following the 1995 parliamentary elections. The President and his clients in Parliament tried to deflect criticism with the rhetoric of free and fair elections, but their words rang hollow after the presidential election the following September. Ter-Petrossyan ran against the country's first Prime Minister, Vazgen Manukyan, receiving fifty-one percent of the vote and narrowly avoiding a runoff that would likely have given Manukyan the job. Both candidates preemptively claimed victory shortly after the election and urged the international community to validate their positions. International observers responded by citing evidence from both sides of "vote tampering, and very serious violations of the election law that [raised] questions about the integrity of the election process and [called into] question the results."[19] Since neither candidate got what he wanted through legal means, three days later a mob of Manukyan's supporters stormed the Parliament building, "broke through the ranks of its defenders and gained entry into the building, where they beat up Speaker Babgen Araktsyan and his deputy. In response, President Ter-Petrossyan banned rallies, ordered opposition parties closed and called in troops and armored personnel carriers to quell the riot and disperse the crowd."[20] Rather than adding support to his cause, the President's strong-arm tactics dramatically increased popular and elite discontent with the election results. Violence eventually broke out between government officials as Parliamentarians took it upon themselves to make their opinions known by "pummeling each other, with opposition legislators getting the worse of the encounter, while televisions broadcast the edifying spectacle to the nation."[21]

To mollify mounting Western criticism for these actions, the President scapegoated his political ally, Prime Minister Hrant Bagratyan, and replaced him with a more familiar face to the West, Armen Sargsyan who had previously served as ambassador to several European countries as well as Head of Mission to the European Community. Under the President's direction, the new Prime Minister continued the work begun by Bagratyan and strove for greater integration into Europe, most importantly entrance into the Council of Europe.

Elite competition had not yet spilled over to the rules of the game, remaining largely confined to disagreements between individual rivals and their political parties. As a result, and as long as the rules remained inviolable, compliance progressed with safety improvements at Metsamor, and the regulatory agency fulfilled a growing list of oversights and recommendations as well.[22] Price stabilization in the energy sector and restructuring of the Energy Ministry to enable more effective oversight of state-run enterprises and increase privatization of non-nuclear production facilities also added to nuclear safety compliance.[23] Ter-Petrossyan supported those costly policies because the advantages of compliance overshadowed the disadvantages due in part to the domestic challenges he faced, as well as the growing rift with Russia over his pro-EU leanings, thereby making the costs of rejecting the West too high. However, several changes began to undermine both the President's position as the head of the informal patronage network and the constitutional procedures governing the formal rules of the game, resulting in lower Armenian compliance than would have been possible otherwise.

Previously, Ter-Petrossyan's clients in the defense, economic, and security ministries had not challenged his authority over the informal rules of the game. In the wake of the worsening institutional and party divides borne out in graphic detail to the electorate, once long-time supports began to turn on the President as he continued to court the West in hopes of greater integration into European institutions. The informal rules of the game began to shift as members of the President's party followed Defense Minister Vazgen Sargsyan in criticizing Ter-Petrossyan for making conciliatory steps towards Azerbaijan.[24] Soon thereafter presidential patronage became less important as Sargsyan set up rival clans and the political costs of defection from the President decreased.[25]

In response to the threat, the President tried to boost his support among veterans and capitalize on nationalist sentiment by appointing former Nagorno-Karabakh leader, Robert Kocharyan to replace Armen Sargsyan as Prime Minister. Kocharyan was extremely popular at the time and the most outwardly Western candidate due to his fundraising in the United States. Unfortunately for Ter-Petrossyan, the move ultimately failed to win him the elite support he had begun to lose following his reelection. In what would have lasting impact on the informal rules of the game and Armenian nuclear compliance for years to come, less than three years after what appeared to be his ascendance to power over the political system, the President was forced to resign under pressure from his former allies and political clients.

The bloodless coup revealed the extent of change in the patronage networks governing the informal rules of the game. Rather than maintain the existing hierarchy, Vazgen Sargsyan used his own influence in the defense and internal affairs ministries to remove the President; the long-standing tradition of outgoing leaders appointing their successors no longer held sway. Yet this shift had far more extensive effects on the rules of the game than merely defrocking a President; it replaced the constitutional process of removing the President with a new informal revolution from below. The informal rules had become unstable, and because of the centralization of power in the presidency, the formal rules were under threat as well.

To highlight the destabilization in the rules, the President's resignation triggered constitutional measures to keep the government running, but those procedures assumed parliamentary dismissal of the President after being convicted by the Constitutional Court, not forced resignation without the Court's review. Instead, succession procedures took on an ad hoc approach, furthering the instability in the system. While the Constitution allowed the Prime Minister to step in as acting President and call for early elections within forty days, and an acting President eventually took office, questions grew as to what he would do and how long he would last in office.

Armenian elites began backing away from their compliance commitments as a result of the growing rules instability. In particular, the government sought alternative funding from Russian sources that would greatly decrease the need for Western assistance requiring early closure. Compliance ground to a halt as mounting delays left uncompleted much of the safety work begun years earlier, and nuclear regulators did not receive the necessary funds to do their jobs in accordance with Armenia's international commitments. Short-term thinking came to dominate the system as Armenia's elites jockeyed for position under greatly destabilized rules of the game, and they looked for ways out of their agreement to decommission Metsamor in 2004, despite Western assurances to hold back their full assistance if that happened.[26]

The final stage begins

In that context, acting-President Robert Kocharyan called for early elections in March 1998 and ran as the incumbent against Karen Demirchyan, the last communist leader of Armenia, but the elections suffered from numerous instances of fraudulent practices. One international observer cited false ballot boxes, some with secret compartments containing stacks of Kocharyan votes, despite the obviousness of the fraud given the ballots pristine, untouched appearance compared to normal ballots.[27] Whether directly a result of election fraud, or simply aided by it, Kocharyan decisively defeated Demirchyan in the run-off election. However, further evidencing the loss of predictable power transfer mechanisms, Demirchyan and his group of supporters followed the precedent of the 1996 election and refused to concede. While he eventually gave in, that aspect of the rules of the game had become more uncertain compared to previous elections.

Despite his election victory, presidential power under Kocharyan had greatly diminished and the formal rules gave President Kocharyan more de jure authority than he effectively possessed. Whereas Ter-Petrossyan could rely on the loyalty of his key clients throughout most of his tenure, the new President appeared weak and ineffectual compared to the new centers of power in government, one of which belonged to Demirchyan. Making matters worse for the new President, his attempted rapprochement with Azerbaijan cost him vital support from the Nagorno-Karabakh veterans headed by Vazgen Sargsyan. Kocharyan incurred the wrath of the original coup-plotters, and Sargsyan's supporters in Parliament called on the President to dissolve the body and hold early elections in October 1998. The President's rivals did not need to force him to resign as with Ter-Petrossyan since the elections would likely have given Sargsyan a sufficient parliamentary mandate to make constitutional changes in his favor. Surprisingly, the President managed to resist their efforts and stuck to the schedule of spring 1999. One can reasonably assume that Sargsyan bided his time out of a belief in the inevitability of his rise to power: he effectively controlled Parliament, Kocharyan did not have the support of the power ministries, and the President's refusal only weakened his position vis-à-vis rival political elites.

Making matters worse for the President, Sargsyan allied with Demirchyan shortly before the May 1999 parliamentary elections to expand their margin of victory. Sargsyan controlled the new patronage system and Demirchyan had widespread popular support.[28] The outcome of the election would either bring stability to the political system through a constitutional reduction of presidential authority that would formalize its already weakened position, or it would keep the fractured balance of power in place until the next presidential election. In the midst of this uncertainty, nuclear officials practically halted all compliance work in anticipation of the election.[29]

Sargsyan and Demirchyan's Unity bloc capitalized on the President's declining image at home and won a sweeping victory to gain a parliamentary majority in May 1999. The President had little choice but to nominate Sargsyan as the new Prime Minister; Demirchyan became the parliamentary speaker, and the new Prime Minister moved his supporters into key government positions. The informal rules had already changed and Sargsyan was in a position to change the formal rules as well.

Uncertainty was actually higher at this point than during the original decline of presidential power in February 1998 because the intervening period had not made official the changes to the formal rules even though the informal rules had changed. This middle ground created contradictory incentives and pressures on elites since the post-May 1999 period removed the earlier formal stability and threw the entire system into confusion. As if to prove the point, President Kocharyan seemed to offer little resistance and appeared almost ineffectual.[30] He had originally counted on the support of veterans before his election, but Prime Minister Sargsyan treated the President as his subordinate after gaining office in May 1999. Given their newfound power, Sargsyan and his supporters could easily create a new center of institutional authority in the Prime Minister, one either at odds with the constitution or that transplanted constitutional powers directly to the Prime Minister. The former choice would have stabilized the informal rules only, while the latter would have affected both formal and informal rules. However, fateful events prevented either from happening and gave the President a chance to bring constitutional and patronage power back under his control.

The opportunity came in October 1999 when armed gunmen burst into the Parliament building and murdered Prime Minister Sargsyan, Speaker Demirchyan, and several other ministers of Parliament.[31] Immediately following the assassinations, Kocharyan turned over the investigation to military and internal security personnel, but Parliamentarians started talking about the President's possible role despite his protestations of innocence. These rumors found listening ears as a few months earlier Sargsyan and Kocharyan had begun publically to show their personal animosity toward each other.[32]

The removal of a parliamentary threat to presidential power and a threat to his patronage network from rival elites did not immediately translate to a resurgence of Kocharyan's power. When the Unity Bloc demanded that he appoint Vazgen Sargsyan's brother, Aram as the replacement Prime Minister, the President acquiesced and appeared to bide his time. Kocharyan did not have to wait long as Aram tried to assume his brother's mantle, but with little political experience and even less power among the military and business owners, could not hold the party bloc together. Factional splits arose over appointments to the defense and internal security ministries, as well as the thorny problem of what to do about the President. As a result, an attempt to impeach the President in April 2000 failed and the fragile Unity Bloc fell apart in the wake of the debacle. Its collapse swung the balance of power decisively in favor of the President, who made his move seven months after the assassinations when in May 2000, Sargsyan's Republic Party split and the President sacked Aram, naming a compliant Andranik Markaryan to replace him.[33]

Kocharyan had won the battle but the political war would continue for years to come because presidential power over the patronage hierarchies had been greatly diminished with Ter-Petrossyan's resignation; the entire system had fractured even further with the assassination of the Prime Minister. If Vazgen Sargsyan had been able to consolidate his control over the formal rules, the system could have stabilized despite the weakening of the President's constitutional powers.

This might have led to a resumption of the progress toward high compliance as governance became more certain, thereby enabling elites to have longer-term perspectives. Unfortunately, the October 1999 assassinations were so disruptive that no one really knew who was in charge or how they would be chosen in the future.

Metsamor in the middle

Informal delineations of power regarding Metsamor had already become unclear prior to Unity's collapse, when the President initially issued a decree to approve employment status for workers at Metsamor, to which Parliament responded by claiming its own oversight in the "Law on the Safe Use of Nuclear Energy for Peaceful Purposes." Both were binding documents that did not officially usurp authority from the other branch of government, but rather indicated competing claims to Metsamor as part of different patronage networks. After the dissolution of the Unity bloc, Kocharyan reasserted presidential preeminence over the nuclear regulatory agency (Decree N-385, June 22, 2000) and energy policy as a whole by using his supporters in the National Assembly to amend the law "On Energy of the Republic of Armenia, March 2001." Both established clear autonomy for regulators by separating them from parliamentary control.

However, despite this progress toward fulfilling the requirements in the CNS, uncertainty about the political system compounded worries about the country's dire energy crisis, and Metsamor once again became a popular rallying cry and among political elites beholden to the electorate. No one wanted to return to the dark days after independence, and nuclear power offered a glimmer of hope out of the painful past.[34] Worse still for those in power, closing it would invite massive popular protest that could hasten an end for any elite's play in the political game.[35] As a result, political costs eventually overwhelmed the commitment to decommission Metsamor before the end of its design lifetime, and high compliance was no longer a possibility for Armenia.

To reiterate the new limitations on compliance, while safety work resumed under ANRA's oversight, the President's Energy Minister, Karen Galustyan, stated officially that Metsamor would not be decommissioned anytime before 2008.[36] Galustyan later said the plant would operate beyond its thirty-year design lifetime to compensate for the time it had not operated in the early 1990s.[37] The European Commission and the EBRD countered by offering loans to cover the costs for developing alternative energy supplies if Armenia stayed on track, but both offers were rejected. European Commission officials then proposed a new timeline since the Armenians could not keep to the original schedule without viable non-nuclear alternatives – the Lake Sevan hydroelectric plant could no longer operate beyond capacity and the country lacked reliable fossil fuel supplies.[38] In the end, the European Commission amended the terms of the PCA with new shutdown dates scheduled for 2006–2007.[39]

No other country received a new term for high compliance, but even these new dates remained questionable. In May 2002 Aram Gevorgyan, head of the atomic energy department of the Energy Ministry, urged Western donors not to force Armenia to close Metsamor in 2006. Four months later another member of the Energy Ministry, Nikolay Grigoryan, stated that the plant could safely run until at least 2016.[40] Government policy followed and in August 2005 President Kocharyan stated that he never felt bound by the 2004 deadline.[41] He was facing a crucial constitutional referendum in November and could not afford Metsamor being used against him.

Aram Sargsyan and Stepan Demirchyan (son of slain Parliament Speaker Karen Demirchyan) had previously challenged him in the 2003 presidential election, drawing support from business leaders, MPs, and citizen groups opposed to the President.[42] The President defeated Demirchyan in the second-round with sixty-seven percent of the vote, but the results were once again widely disputed by the losers, soundly criticized by international observers, and rejected by opposition parties in Parliament. The Constitutional Court upheld the results but recommended a vote of confidence in the President, the first time it had stepped out of the President's control. The vote did not take place since parliamentary elections followed in May 2003 and the President's supporters won a majority of seats. Opposition parties claimed fraud but again to no avail.

President Kocharyan continued his attempt to minimize alternative bases of power by pressing for a constitutional referendum that would give his control over the government the appearance of being more democratic. The proposal would grant Parliament more oversight over the government and the judiciary, while maintaining emergency powers in case the President's supporters lost in future elections. This would also remove the problematic image of de jure presidential control of the Constitutional Court while maintaining de facto control since it had proven unreliable.

Sargsyan led opposition parties in boycotting the vote, blasting the President and his patronage network, and going so far as to call for total revision in the rules themselves based on the arbitrariness of the system. He stated emphatically that the President and his group "live by other laws, today they say that the situation in the country is bad . . . and the old constitution is to blame for it, [and] if you change it, the situation will change. It won't, the same prefect, the same policeman, the same mayor, nothing will change. Things will only change in this country through revolution."[43] National Union-Democratic Union Party leader Vazgen Manukyan echoed Sargsyan's cynicism arguing that "with the new changes there will be a redistribution of powers among the President, the National Assembly, and the government. That is, the ruling feudal-oligarchic system is deciding among themselves which power will be divided in which way and with this division the system becomes more stable."[44] Former Minister Raffi Hovannisyan went further to claim Kocharyan and his cronies in Parliament were a "regime that supports thieves, murderers and corrupt individuals and enjoys their support."[45] Of primary concern was Article 56 that would give the President immunity similar to sitting MPs (Article 66), and changes to Article 53 that would allow the President to remain in office indefinitely during times of martial law or state emergency since he would not need any outside approval or oversight to issue such a decree.

Despite the damage to open, free, and fair elections, as well as the potential for long-term autocracy in the President's plan, some changes appeared to resolve the unbalanced constitutional powers favoring the President and to ensure greater stability in the formal rules. In particular, the new system offered parliamentary approval of the President's nominee for Prime Minister, the removal of presidential oversight of the Justice Council, and the popular election of the Mayor of Yerevan in place of presidential appointment. None of these would have lessened the conflicts between the President's party and his opponents in Parliament, but the rules would have clarified the division of power between the institutions of government. Unfortunately, the ongoing battle between rival elites and their clients in Parliament and local government showed no signs of stopping before the vote, and political violence remained a distinct possibility.[46]

Kocharyan needed one-third of all registered voters (750 thousand) to approve the draft, but opposition groups hoped turnout would be low enough to deny the President any opportunity to rig the election. However, the vote tally showed 1.5 million out of 2.3 million eligible voters turned out and ninety-four percent voted in favor of the referendum even though members from the European Council's parliamentary Assembly and Congress of Local and Regional Authorities observation teams cited numerous instances of ballot box stuffing and illegal voting.[47] The Armenian group "It's Your Choice" also reported violations of voter secrecy, the presence of unauthorized personnel in polling stations, violations of citizen's right to vote, and abuse of observers.[48] In addition, turnout around the country was believed to be considerably lower than the official numbers, thereby calling into question Kocharyan's victory. As a result, Sargsyan and Demirchyan called for mass protests and horn honking in Yerevan to protest the results, drawing a swift response from police who warned of reprisals against anyone violating public order.[49] Despite the criticisms though, Kocharyan prevailed and the new system began once again to bring stability to the formal rules of the game even as the informal rules remained unstable due to the entrenched rival patronage networks.

Conclusion

Rules stability and compliance in Armenia had three distinct time periods beginning with the signature of the PCA with the European Commission and concluding with Metsamor operating beyond 2007. Time period 1 showed increasing stability under Ter-Petrossyan leading to increasing compliance from 1996 to early 1998. The uncertainty of the interim time period 2, from February 1998 to May 2000, made governance difficult and called into question the formal division of power between the Prime Minister and the President. Elites were hard pressed to make any long-term decisions to resolve the country's problems, let alone address nuclear safety compliance. As a result, they held onto Metsamor even though it meant violating signed international agreements and jeopardizing future cooperation with the West. Low rules stability produced short-term thinking that led to a retreat into low compliance.

Time period 3 began after Kocharyan reestablished formal presidential power in May 2000 and continued until Armenia officially missed the 2007 decommissioning deadline. It can be best described as having medium rules stability since the informal rules remained unstable with competing patronage networks vying for influence within the presidential administration and against it in Parliament.

In contrast, even though the Constitution changed again after a November 2005 referendum, the loss of some executive privilege has not limited the President's control over the government, a key formal power. He can still dismiss the National Assembly if members refuse his choice for Prime Minister, and his unilateral authority to declare martial law and remain in office indefinitely maintains de facto presidential preeminence.[50]

Compliance followed the increase in rules stability in 2000 but capped at medium levels. Safety measures are currently up to date and the nuclear regulator has structural independence from the state utility and plant management.[51] However, since Metsamor did not close by 2007, the last date presented by the European Commission to enable high compliance, medium compliance was as high as they could ever reach. In addition, without reliable oil and gas alternatives to Metsamor, and the unfeasibility of relying on the Lake Sevan hydroelectric plant again, Armenia's dependence on nuclear power has not diminished meaning the country will likely continue to rely on it for decades to come.[52]

Table 3.1 Armenia Time 1 (1991–February 1998)

Rules Stability	*Clarity*	*Predictability*
Formal	Yes	Yes
Informal	Yes	Yes
Conclusion: Informal rules operated based on predictable patronage networks determining political and economic privilege. These combined with formal rules that favored the concentration of power at the top of the patronage hierarchy. **Moving toward HIGH STABILITY**.		
Compliance	• Safety upgrades • Creation of ANRA • 1996 PCA commitment to early decommissioning	
Conclusion: Armenia was taking the necessary steps and **moving toward HIGH COMPLIANCE**.		

Table 3.2 Armenia Time 2 (February 1998–May 2000)

Rules Stability	*Clarity*	*Predictability*
Formal	No	No
Informal	No	No
Conclusion: Informal revolution from below created an alternative patronage network that threatened the position of the presidency atop the hierarchy. This produced uncertainty about the formal rules as well, resulting in **LOW STABILITY**.		
Compliance	• Considerable delays in safety work • No additional legislation to develop ANRAs financial and bureaucratic autonomy and competing claims to its oversight • 1996 PCA commitment to early decommissioning called into question at best, rejected at worst	
Conclusion: The country still produced **LOW COMPLIANCE** since some safety work had occurred earlier in the period.		

Table 3.3 Armenia Time 3 (May 2000–2007)

Rules Stability	*Clarity*	*Predictability*
Formal	Yes	Yes*
Informal	No	No
Conclusion: Multiple competing patronage networks vie for control of Parliament and government ministries. *Formal power has consolidated under presidential control despite some constitutional prerogatives going to Parliament producing **MEDIUM STABILITY**.		
Compliance	• Safety upgrades • ANRA given more autonomy • 1996 PCA commitment to early decommissioning rejected and operation likely to continue until 2016	
Conclusion: ANRA has sufficient autonomy even though it still lacks sufficient funds. International assistance helps the Agency to ensure safe operation showing **MEDIUM COMPLIANCE**.		

Contrasting these variations, safety culture among ANRA personnel has remained primarily risk averse despite changes in compliance, and its safety culture guides daily operations given the regulator's effective oversight of the plant. In addition, international incentives have not varied in nature even if the timeframe changed to allow for a longer decommissioning process.[53] Western donors never offered enough money according the Armenians, and neither the amount nor Armenia's need for it changed. In contrast, compliance increased under Ter-Petrossyan, moved back down to low in the years of chaos, only to rebound to medium levels after May 2000. As a result, rules stability explains compliance behavior more effectively than the two primary counter arguments of changing safety culture and international incentives.

In conclusion, while Armenia's political system has not lost any of its verbal venom since the 1999 political assassinations, enduring bitterly disputed presidential elections in 2008 and 2013, it survived for almost a decade before the next example of that kind of violence occurred with the attempted assassination of presidential candidate Paruyr Hayrikyan in February 2013. The incident raises the specter of a return to the troubling days of unstable rules, instability that both permitted and perhaps fostered violence in the political arena, but those threats may not materialize. In their place, the rules of the game have achieved medium stability due to stable formal rules and unstable patronage networks initiated with Ter-Petrossyan's ouster. The former President's attempted return to office in 2008 and bitter dispute with the election results did not throw the system into chaos even though it signaled a possible shift in patronage alliances. It appears likely that medium rules stability will persist for the foreseeable future. As a result, Metsamor will continue to operate without a sufficient containment structure, but at least guided by a nuclear regulatory agency that has effectively implemented necessary safety upgrades and works to ensure safe operation. In the final analysis of what was possible after 2007, medium compliance is still better than the alternatives because the risks of disaster have diminished.

Notes

1 Jivan Tabibyan, in discussion with the author, September 2002.
2 Daly (2013).
3 See Dudwick (1997) for more on the different elites, their party platforms, and the general tone of the government during this period. Way (2006) carries that research into the next decade.
4 Herzig (1999) gives more details on the types of challenges facing the country during the transition.
5 *Nucleonics Week*, January 9, 1992: 1.
6 Ishiyama and Kennedy (2001) provide in-depth analysis of these influences on the political system at large.
7 One of the chief weaknesses of Armenian democracy was the lack of an effective civil society as the populace grew more disillusioned with elected officials and political institutions after 1995 (Dudwick, 1997; Karatnycky, 1998).
8 Market reforms included the establishment of a new currency based on strict IMF guidelines, tax reforms, privatization of many state enterprises, deficit reductions, and the elimination of subsidies on consumer goods (Commission on Security and Cooperation in Europe, 1995: 3).
9 This centralization of power in one person stood at odds with liberal democracy, but it was more stable than the hyper-presidential systems in other post-communist countries. See Ishiyama and Kennedy (2001) for solid comparisons.
10 Article 6 made sacrosanct these powers by locating supreme legal sovereignty in the constitution.
11 *Nucleonics Week*, November 5, 1992: 12.
12 *Nucleonics Week*, February 4, 1993: 4.
13 *Nucleonics Week*, January 9, 1992: 1.
14 The Russian Ministry of Atomic Energy advised the Armenians to restart immediately, informing Western nuclear officials of their counsel as well (Armen Abagyan, in discussion with the author, September 2002).
15 *Nucleonics Week*, September 15, 1994: 3.
16 Former US Senior Nuclear Safety Coordinator, Carol Kessler, confirmed that the regulatory staff held plant personnel to very high standards of conduct, and pursued safety with as many resources as they received (Carol Kessler, in discussion with the author, personal interviews, 2002).
17 Carol Kessler confirmed to me that she personally secured a commitment by Prime Minister Bagratyan that Metsamor would be shut in 2004 because Armenia wanted to become part of the West in the future, and they recognized early closure was the price of doing so (Ibid.).
18 NATO Parliamentary Assembly (1999): 6.
19 Commission on Security and Cooperation in Europe (1996): 3.
20 Ibid., 15.
21 Ibid.
22 See International Atomic Energy Agency (1999).
23 As evidenced in two laws put into force: the Law on Energy of the Republic of Armenia and the Law on the Safe Use of Nuclear Energy for Peaceful Purposes.
24 Danielyan (1998) discusses the significance of Nagorno-Karabakh in greater detail here.
25 See Herzig (1999) for more details, but one indication was that future President Robert Kocharyan served as a key member of Sargsyan's faction as both had fought in Nagorno-Karabakh, Sargsyan as the leader of defense forces and Kocharyan as the former President of the republic.
26 This behavior did not represent total non-compliance despite the slowdown. Some important steps had been taken to improve safety, and while lacking sufficient funding and resources, the regulator possessed a degree of independence from other bureaucracies.
27 See Commission on Security and Cooperation in Europe (1998).

28 Inter-parliamentary Union (May 30, 1999).
29 Ashot Martirosyan, in discussion with the author, September 2002.
30 Commission on Security and Cooperation in Europe (1999): 13 confirms this assessment.
31 Former journalist Nairi Hoonanyan and a few other disgruntled citizens carried out the slaughter.
32 Freedom House (2001).
33 While Vazgen Sargsyan had been the military leader of Nagorno-Karabakh, many of the veterans sided with the President without Sargsyan as a counterweight.
34 Ambassador Tabibyan told me that no one knew what would happen next during those chaotic times, and that Metsamor was one of the only sure things they could count on (Jivan Tabibyan, in discussion with the author, September 2002).
35 Many of the deaths after the 1988 earthquake came from freezing after houses and heating systems collapsed. Those memories lingered with the fear of closing Metsamor and losing its power generation.
36 *Armenian International Magazine* (2000): 30.
37 Bellona, October 5, 2000.
38 Personal correspondence with David Sycamore, Foratom Institutional Affairs Director and formerly at the European Union, December 2002.
39 European Community (a, b, c) give evidence to that fact.
40 *Nucleonics Week*, May 23, 2002: 13.
41 Danielyan (2005)
42 Danielyan (2004) details the types of groups and interests involved in the opposition movement.
43 Ishkhanian (2005): 120.
44 Ibid.
45 Avoyan (2005).
46 Violence had been used before during contested elections and the American Embassy in Yerevan issued a warning it could happen again (Ishkhanyan, 2005).
47 Buniatian and Saghabalian (2005) provide examples from the commonplace to the surreal.
48 Organization for Security and Cooperation in Europe (2005).
49 See Danielyan and Kalantarian (2005).
50 *NIT* Rule of Law scores from 1997–2003 confirm the rise and fall in the political system. I do not rely on them exclusively, but they offer corroborating evidence for the general trends. Ratings use a scale of 1 to 7: 1 represents the highest level of democratic progress and 7 the lowest. A change of .25 or more is considered highly significant.

NIT Ratings	*1997*	*1998*	*1999–2000*	*2001*	*2002*	*2003*
Rule of Law	4.75	5.00	5.38	5.38	5.38	5.38

51 These latter steps are important parts of medium compliance so I give the Armenians credit for the effort even if they have not fulfilled every detail to the last degree.
52 Retrofitting and refueling can extend Metsamor's lifetime for many more years, as advocated by Russian advisors.
53 See the following for examples of consistency: International Atomic Energy Agency (2001); "Armenia Safety Culture" (2007); "Soviet Nuclear Legacy," (2007); Organization for Economic Cooperation and Development/Nuclear Energy Agency (1999, 2000a, 2000b); US DOE (1996–2002).

References

"Armenia Safety Culture." Accessed January 30 2007. www.amecnnc.com/pdf/RS21%20-%20Safety%20Culture.pdf.

Armenian International Magazine. "Nuclear Power Plant to Remain Operational Beyond 2004." 11 (November 2000).

Avoyan, Shakeh. "Lashing Language by Hovannisian: Former FM Joins Boycott against 'Ruling Clique'." Accessed December 1, 2005. http://armenianow.com/news/6038/lashing_language_by_hovannisian_fo.

Bellona. "Armenian NPP to Receive Free Fuel from Russia," October 14, 1997; "Armenian NPP Back in Operation," December 9, 1997; "Armenia Will Build a New NPP," January 15, 1998; "2004 Closure of Metsamor NPP Ruled Out," October 5, 2000; "Armenia Owes Russia $36 Million for Fresh Nuclear Fuel," December 4, 2000; "The Armenian NPP to be Launched Again," October 22, 2001; "Armenia Needs $8 Million for Nuclear Fuel," June 20, 2002.

Buniatian, Heghine and Anna Saghabalian. 2005. "Armenian Observers Slam 'Undemocratic' Referendum." *RFE/RL Armenia Report.* 29 November.

Commission on Security and Cooperation in Europe. "Armenia's Parliamentary Election and Constitutional Referendum, July 5, 1995."

——. "Armenian Presidential Election, September 22, 1996."

——. "Report on Armenia's Presidential Election, March 16 and 30, 1998."

——. "Report on Armenia's parliamentary Election, May 30, 1999."

International Atomic Energy Agency (IAEA). Convention on Nuclear Safety, 1994.

Daly, John. "Armenia's Metsamor NPP, Built near Fault Line, Gets 10 Year Life Extension." Accessed September 30, 2013. http://oilprice.com/Alternative Energy/Nuclear-Power/Armenias-Metsamor-NPP-Built-Near-Fault-Line-Gets-10-Year-Life-Extension.html.

Danielyan, Emil. "Armenia: Change in Leadership without Political Reform." In *Holding the Course: Annual Survey of Eastern Europe and the Former Soviet Union*, edited by Peter Rutland, Chapter 89. London: M.E. Sharpe, 1998.

——. Eurasianet.org. Accessed regularly 2004. www.eurasianet.org/taxonomy/term/1404?page=6.

——. "Armenian Nuclear Plant to Function for another Decade." *Eurasia Daily Monitor*, 8 (August 2005).

Danielyan, Emil and Karine Kalantarian. "Opposition Condemns 'Unprecedented' Fraud, Vows More Protests," 2005. Accessed January 15, 2006. www.armeniapedia.org/wiki/Constitutional_Referendum_of_2005.

Dudwick, Nora. "Political Transformations in Postcommunist Armenia: Images and Realities." In *Conflict, Cleavage, and Change in Central Asia and the Caucasus*, edited by Karen Dawisha and Bruce Parrott, chapter 3. New York: Cambridge University Press, 1997.

European Community (a). "TACIS Nuclear Safety: Action Program 2000 – Republic of Armenia."

——. (b). "Technical Assistance Program: Action Program 2000 – Republic of Armenia."

——. (c). "Technical Assistance Program: 2000–2003 Indicative Program – Republic of Armenia."

Freedom House. 2001. Accessed regularly 2002–2012. www.freedomhouse.org/report-types/freedom-world.

Herzig, Edmund. *The New Caucasus: Armenia, Azerbaijan, and Georgia.* London: Pinter, 1999.

International Atomic Energy Agency. *Report of the International Conference on Strengthening Nuclear Safety in Eastern Europe*. Vienna, Austria, June 1999.

——. *Second Review Meeting of the Contracting Parties – Convention on Nuclear Safety*. Vienna, Austria, April 2001.

Inter-parliamentary Union. Accessed regularly 2002–2012. www.ipu.org/parline-e/reports/2013_E.htm.

Ishiyama, John and Ryan Kennedy. "Superpresidentialism and Political Party Development in Russia, Ukraine, Armenia, and Kyrgyzstan." *Europe-Asia Studies* 53, no. 8 (2001): 1177–1191.

Ishkanian, Armine. "Diaspora and Global Civil Society: The Impact of Transnational Diasporic Activism on Armenia's Post-Soviet Transition." In *Central Asia and the Caucasus: Transnationalism and Diaspora*, edited by Touraj Atabaki and Sanjyot Mehendale, Chapter 6. London: Routledge, 2005.

Karatnycky, Adrian, Alexander Motyl and Boris Shor, eds. Nations in Transit 1997: Civil Society, Democracy and Markets in East Central Europe and the Newly Independent States. New Brunswick: Transaction Publications, 1998.

North Atlantic Treaty Organization Parliamentary Assembly. "Nuclear Safety in Central and Eastern Europe," (AU 119): STC (01) 4. 1999.

Nucleonics Week, January 9, 1992: 1; November 5, 1992: 12; February 4, 1993: 4; September 15, 1994: 3; May 23, 2002: 13.

Organization for Economic Cooperation and Development/Nuclear Energy Agency. *Assuring Future Nuclear Safety Competencies*. Paris, France, 2000.

——. *Regulatory Response: Strategies for Safety Culture Problems*. Paris, France, 2000.

——. *Russian Minatom Nuclear Safety Research Strategic Plan – An International Review*. Paris, France, 1999.

Organization for Security and Cooperation in Europe, Office for Democratic Institutions and Human Rights. "Republic of Armenia Constitutional Referendum 27 November 2005 – Needs Assessment Mission Report October 24–25." Warsaw, November 10, 2005.

"Soviet Nuclear Legacy." Accessed January 30 2007. www.dti.gov.uk/energy/environment/soviet-nuclear-legacy/programme/nsp/nsp-armenia-profile-safety2004/page16701.html.

US Department of Energy, Office of International Nuclear Safety and Cooperation. *Monthly Activity Reports for Improving the Safety of Soviet-designed Nuclear Power Plants*. Washington, DC, 1996–2002.

Way, Lucan. "Pigs, Wolves and the Evolution of Post-Soviet Competitive Authoritarianism, 1992–2005." *Center on Democracy, Development, and the Rule of Law*. Working Paper 62 (June 2006): 1–64.

Additional Reading

Armenia Liberty. "Opposition Claims 'Purges' in Election Bodies," March 4, 2002; "Governing Party to Go It Alone in Parliamentary Polls," March 4, 2002; "Candidates in Armenian Presidential Election," March 4, 2002.

Cooperman, Alan. "The Next Chernobyl? Armenia's Metzamor Nuclear Power Plant is online. Hold Your Breath." *US News and World Report* 120, no. 17 (April 29, 1996): 46–49.

European Union and the Republic of Armenia: Partnership and Cooperation Agreement, 1996.

ITAR/TASS. "Mothballing of Armenian Nuclear Power Plant Premature," May 10, 2001.

Nuclear Energy Institute. *Source Book on Soviet-Designed Nuclear Power Plants, Fourth Edition*. Washington, DC, 1997.

——. *Source Book on Soviet-Designed Nuclear Power Plants, Fifth Edition*. Washington, DC, 1998.

Radio Free Europe/Radio Liberty "Armenia: Prime Minister, Three Others Killed in Parliament Shooting," October 27, 1999; "Brother Named Successor of Armenia's Slain PM," November 3, 1999; "Kocahrian Presents New PM to Cabinet, Says Most Ministers Should Stay On," November 4, 1999; "Former PM and Brother of Slain PM Joins Presidential Race," December 4, 2002.

Wilson, Richard. "Armenian Nuclear Plant Report." *Andre Sakharov Foundation* (July 25, 1995).

4 Lithuania reluctant but resolved to obey

"Ignalina is safe. We continue to implement safety measures on unit 1 and especially unit 2, but the political decision is done. The reactor is OK, but Lithuania is pursuing EU membership. This is the price to pay."

—Saulius Kutas, Chairman, Lithuanian State Nuclear Power Safety Inspectorate (VATESI), September 2002

Introduction

The Ignalina nuclear power plant was the largest in the world with nearly 3,000 megawatts of generating capacity. It provided over seventy percent of Lithuania's electricity needs and exported several billion kilowatt hours to Poland, Estonia, Belarus, and Russia. The massive RBMK reactors underwent extensive safety upgrades and modifications over two decades of cooperation with the West, thereby enabling it to generate considerable revenue for the state in the process. Lithuania desperately needed that money to solidify its growing economic recovery, but unfortunately, the Ignalina power plant was a ticking time bomb given the design flaws it shared with Chernobyl.

Despite the risk of nuclear disaster, plant management, regulators, and politicians stressed their confidence in the plant's safety even though two considerable obstacles stood in the way of Ignalina's continued operation. First, a 1994 Grant Agreement with the EBRD explicitly called for both reactors to shut down operations by 2009, and second, the European Union demanded Lithuania begin decommissioning before the country joined in 2004. Lithuanian politicians regularly expressed dismay at being forced to close Ignalina, but they still fully complied with their international commitments. Even though they deeply resented the cost of membership, elites fulfilled their requirements and decommissioned both Ignalina units on time, years before the end of their official lifetimes. Stable rules of the game enabled that decision.

Lithuania enjoyed stable political rules during the fifteen years between the signing of the EBRD Grant Agreement and final shutdown in 2009. The formal structures of government established in the constitution did not undergo any fundamental alterations as seen in predictable power transfer mechanisms and unchallenged delegations of authority between Parliament and the President.

A few isolated efforts to bring more power to the presidency arose, but none proved lasting, and the system emerged unchanged with even more deeply entrenched rules.

The informal procedures of normal day-to-day governance also enjoyed high stability. Political offices transferred smoothly and predictably, governments alternated according to predictable election cycles, and only once did the party in power change before the completion of its term – a minority coalition lost key support from smaller parties and the election winners took over the government. Additionally, bureaucratic responsibilities did not change despite successive changes in government, with stable delegations of authority between ministries as well. As a result, stability in both sets of rules enabled elites to pursue the long-term benefits of compliance despite the enormous costs they required.

Troubled beginning, productive lifetime

Ignalina came online three years before Chernobyl exploded. Amazingly, despite the dangers of a similar accident given its RBMK design, management did not implement quality-assurance control programs after the disaster in Ukraine. For nearly a decade the fire control systems were completely inadequate with slow-closing wooden doors leading to the control room, flammable paint and floor tiles covering the inside of the facility, and the ventilation system connecting most of the rooms in the building, including those with safety monitoring and control systems. Early assessments by the IAEA also revealed faulty welds at the tops of pressure valves at the second unit, greatly increasing the risk of accidents getting out of hand. Director of special projects for the Swedish Nuclear Power Inspectorate, Jan Nistad, commented on the problem in early 1993, saying that "if the top breaks completely, you will have a missile going through the roof. A break would also create a local power increase in the core because the boiling water coolant in the fuel channel around the pressure tube would be drained."[1] As the Cold War ended and relations with the West increased, scientific exchanges between international and Lithuanian safety experts raised everyone's attention to the myriad problems at Ignalina. Thus by the early 1990s, little doubt remained as to the need for serious safety modifications, upgrades that needed to occur quickly if the plant was to operate for the foreseeable future.

The decision to begin safety modifications took place during the transition from the Soviet Union. Leading the change, the Sajudis independence movement advocated for a law-based, democratic system and received support from the former Communist Party for those goals.[2] Vytautas Landsbergis guided Sajudis through the early transition and constitution formation periods, helping to produce several political parties that governed the country over the coming years.[3] However, as Russia began removing subsidies from Lithuania, the economy collapsed and the electorate turned on Sajudis. In October 1992, reconstructed Communists under the new Democratic Labor Party of Lithuania (LDDP) soundly defeated a majority of Sajudis-backed candidates and gained control of the new unicameral

Parliament. The LDDP government espoused a similar pro-European position as its rivals, but offered greater social assistance to those hurt by marketization and the decline in living conditions. The situation worsened for Sajudis in December as economic woes and infighting led to even larger defeats. The return of the former Communists did not spell an end to democracy though. A referendum in October 1992 had firmly established the constitution, free and fair elections set the standard for future contests, and no political figures of distinction hatched any significant anti-democratic designs.[4]

Over the next four years, the former Communists governed the country by controlling Parliament, the government, and the presidency. This enabled political elites a measure of peace to finish resolving some of the most pressing and challenging issues during the early transition period. The framers chose minimalist strategies rather than negotiate every issue at length. The resulting policies did not call into question the separation of powers or specific power transfer mechanisms, instead giving time for the relative merits of different measures to become clearer over time, and for leaders to resolve their differences through established democratic means rather than the political violence and turmoil plaguing many of the other former communist countries in the region. The strategy worked and politicians habituated to the new environment, producing stable governance during the 1990s and 2000s. The progressive stabilization of the rules of the game enabled increasing compliance as a result.

However, despite real progress in the transition from communism, serious problems lurked in the background regarding Ignalina. Lithuania initially signed and ratified the Vienna Liability Convention in 1993, thereby enabling Western safety companies to work on the plant without fear of legal punishments for accidents, but delays in the implementation of the Convention over the next year slowed even the simplest tasks; equipment and material often sat unpacked outside the plant. Ignalina personnel had already started some of the smaller projects themselves, but serious financial problems plagued the plant throughout the 1990s, limiting their efforts.

Making matters harder, Lithuania depended on Russia for fossil and nuclear fuel shipments, and because Russia maintained a confrontational position toward the break-away republic after 1991, Ignalina faced intermittent fuel supply shortages.[5] In addition, few customers could pay the new electricity bills resulting from the government's marketization policies, making cash-flow a serious problem as well. The two largest customers, the Lithuanian State Utility, Lietuvos Energija, and Belarus, racked up hundreds of millions of dollars in debt to Ignalina as prices rose, leaving safety work increasingly dependent on foreign assistance. The pattern continued for more than a decade, only improving after Lithuania joined the European Union and payments by the state utility increased. Ironically, the plant became most profitable as it moved inexorably toward permanent shutdown.

Despite the changes to Ignalina's financial viability, compliance has consistently improved since Lithuania entered into the Grant Agreement with the EBRD to fund short-term upgrades, then ratified the Convention on Nuclear Safety to improve the status of the nuclear regulator, and finally established the necessary decommissioning procedures to close the plant. Assisting in those efforts, Lithuania already had a functioning regulatory agency that needed fewer changes than its counterparts in the other post-communist nuclear countries; its members also supported the need for safety upgrades from the start of the work. That support still butted up against the Grant Agreement's clear expectations that international assistance required early closure of unit 1 sometime between 1998 and 2004, and 2009 for unit 2. By agreeing to the terms of the Grant Agreement, the Lithuanian government acknowledged the reactors could not be brought to acceptable safety levels given the current level of technology.[6]

The dates had been established based on the lifetime of operation for the fuel channels in the reactor's core. As RBMK fuel channels get older, the gaps between the graphite moderator and the fuel channel begin to close, eventually forcing their replacement. The reactor must cease operations if they are not replaced. VATESI and Western experts calculated the dates for early decommissioning based on the expected gap closure rate for Ignalina. Accordingly, the Grant Agreement stated that the plant could operate beyond 2009 only if VATESI certified the channels' continued operations met Western standards, or if energy demand rose sufficiently to justify rechanneling as the least-cost option. Despite those two exemptions, Bank officials and donor states did not expect either condition to be met, thus requiring Ignalina to close as planned.

Unit 1's 1998 decommissioning had been an ambitious timeframe, and as the deadline approached, plant manager Viktor Shevaldin wrote to the EBRD asking permission to replace the ten worst channels as a way of establishing a baseline for all the channels. Management was under no obligation to ask for this permission since limited replacement fell under the routine maintenance program for an RBMK, and therefore valid according to the Convention on Nuclear Safety. In response, officials at the Nuclear Safety Account (NSA) overseeing the safety work blasted Shevalidin and VATESI for violating the terms of the Grant Agreement, which stated that the nuclear regulator, not plant management, would determine the length of time the fuel channels could operate. NSA director Lars Larrson emphatically stated that any rechanneling constituted lifetime extension and was completely out of the question. The bank's firm resistance may have been more bluster than a defensible interpretation of the agreement as "a source close to the NSA donor's group said he believed replacing only a few channels constituted maintenance, not life extension [and] that the Lithuanian government's decision not to do major rechanneling hadn't changed. But to change individual channels, that's just like doing any other kind of maintenance on any other unit."[7]

Shevaldin read the agreement the same way and proceeded with VATESI's blessing to replace the ten worst fuel channels. In September, Regulatory Chairman Saulius Kutas simply stated that "they were taken out. Nothing [was] done for life extension. We will study them to find out more about what is happening in the reactor. The process will continue with every [maintenance] outage."[8] One week later, bank officials and donors conceded that limited replacement of fuel channels did not constitute lifetime extension for an RBMK reactor. What had appeared to be a clear violation of Lithuania's international commitment was nothing more than routine maintenance. The incident was one of a few cases where on the surface the Lithuanians appeared to be in non-compliance, or at least heading in that direction, but in every instance kept moving toward high compliance as they honored their agreements.

VATESI: A diamond in the rough

VATESI maintained a central role in all safety work and later decommissioning efforts since its inception in 1992. The Convention on Nuclear Safety requires member states to provide sufficient financial resources, personnel, institutional autonomy, and bureaucratic authority to ensure that the regulatory agency determines plant safety, and that the government fulfills its national and international safety requirements. VATESI consistently did its job with limited personnel and initially limited financial resources by utilizing international assistance and making the most of its institutional autonomy and legal authority.[9]

To that end, beginning in 1994, VATESI began a comprehensive Safety Analysis Report (SAR) of the entire plant under the auspices of the Grant Agreement. The original timeline called for VATESI to complete the report by the end of 1995, but doubt overshadowed that date almost immediately. The depressed national economy and limited qualified personnel slowed down the process since VATESI personnel originally came from a relatively small group; most of the Ignalina personnel were Russians trained by the old Soviet Ministry of Atomic Energy. It took time for Lithuanian universities to produce trained nuclear engineers of their own, but between 1998 and 2001 the organization added twenty inspectors, and seven more by the end of 2003.

While regulators worked on the report, the EBRD and the European Commission reminded the Lithuanians of their commitment to close the first Ignalina unit by 1998, although the Agreement allowed for closure as late as 2004. Western officials indicated to VATESI that after completing the SAR, licensing preparations should take one year, at which point VATESI should grant a two-year operating license for unit 1. The timeline set December 1995 for completing the report, December 1996 for the license, and decommissioning the reactor in 1998. The plan was ambitious and did not take into account the personnel and financial limitations facing VATESI. Thus by the time VATESI finally completed the safety report in late January 1997, the whole schedule had been thrown off track. An international panel of Western safety experts criticized VATESI for taking so long, and blamed the problem on insufficient resources.[10] However, the Grant Agreement did not contain a specific timeframe for completing the SAR or the licensing procedure thereafter, and regulators had completed their tasks within a reasonable timeframe. In addition, the Grant Agreement gave a target range for decommissioning unit 1, and VATESI stayed on schedule when it finished the SAR and licensed the plant in August 1999 for two more years of operation. The West had been pressuring Lithuania for the earliest possible date, but the latter date was still within the bounds of the agreement.

Stable rules prevail

The appearance of violations and problems with compliance mirrored the political system, which appeared less stable than it really was. On the positive side, the proportional representation system allowed parliamentary majorities to form based on the actual percentage of votes cast, not arbitrary backroom deals, and party politics determined how the government functioned on a daily basis. Once established after the election, governments followed those practices and established long-standing bureaucratic responsibilities, thus helping to make the informal rules very stable. The constitution reinforced such stable governance by protecting institutional checks and balances. In particular, the President received considerably less power than Parliament, but could still call for new elections and dismiss the Prime Minister with parliamentary approval. The President's status as the only nationally elected official translated to a strong popular mandate with an overwhelming electoral victory, thereby giving him more opportunity to pressure the government to follow his lead. This political influence was not codified in the structural balance of power established within the constitution though; it depended entirely on popular opinion.

Presidential vetoes also had more to do with party politics than the institutional fighting found in less stable post-communist systems. During the early years after independence, the country's first President, Algirdas Brazauskas, had very little need to use his official veto power since his Democratic Labor Party controlled Parliament. He only sent back twenty-three of the nearly one thousand bills passed by the Seimas (Parliament), of which less than ten Parliament overruled. Thus, party politics helped smooth out the early years of transition and further entrench the formal rules of the game. As a result, "though somewhat clumsy, the institutions are well balanced."[11]

However, stable formal and informal rules did not preclude other serious problems from showing up early in the transition from communism. In their first term in office, LDDP could not solve the major economic problems facing the country and corruption charges against key political figures began to grow. Prime Minister Adolfas Slezevicius became the most prominent target.[12] In addition to charges against his cabinet, he personally faced criticism for his December 1995 withdrawal of $30 thousand from his bank the day before it closed. He avoided losses that left thousands of pensioners, business owners and families without access to the meager savings they had amassed in hopes of moderating the worsening economic conditions. "His touching story that he wanted to buy his wife a car for Christmas did not endear him to the mass of small depositors who received no advance warning and whose accounts were frozen after the crash . . . The case along with others suggested that corruption reached right to the top and that the government had lost its credibility."[13] Even though it reached to the heights of government, the nature and scope of corruption was fairly benign compared to other countries. The illegality of the Prime Minister's actions was clear, but the formal and informal rules of the game moderated the damage such actions produced. LDDP suffered embarrassment and lost political legitimacy, but the political system had established mechanisms for dealing with wayward politicians, and thus the rules remained undisturbed by the machinations of a few high-ranking officials.

The most important such mechanism centered on how public opinion shaped government appointments and election results. In Slezevicius' case, widespread outrage eventually forced President Brazauskas to distance himself from the Prime Minister since his primary concern focused on reelection less than a year away. At first, Slezevicius refused to resign. He had already survived a vote of no-confidence in July 1994 so he may have figured he could weather this storm as well. His reluctance only added to public ire though, and charges of corruption, regardless of their veracity, grew with each passing day he remained in power. Finally on February 8, 1996, both ruling and opposition parties in Parliament overwhelmingly voted no-confidence in the Prime Minister and forced Slezevicius to resigned. Since the configuration of party politics remained unchanged in the Seimas, Brazauskas appointed another LDDP member, Laurynas Stankevicius as caretaker in preparation for the October elections. While the change in government resulted from corruption charges and the resulting public furor against it, the day-to-day practice of politics remained unchanged. The President and Prime Minister were both members of the same party, and they kept much of the cabinet from the previous government. The former communists still controlled the legislative agenda, and the mechanisms for power transfer within the government remained unchanged, as did the delegation of power both between and within government institutions. No elites openly challenged the rules; even Slezevicius operated within his legal right to remain in office until Parliament removed him. It may have looked bad, but the Lithuanians followed normal procedure; even when the minority Liberal Union government collapsed in 2001, a new Social Democrat government formed in its place without changes either to the governance procedures or bureaucratic responsibilities for Ignalina.

The formal rules also survived without major revisions despite an attempted challenge by President Brazauskas following LDDP's losses in the 1996 parliamentary election. After his party's defeat, the President found himself politically isolated and forced to nominate a Homeland Union-Lithuanian Conservative (TS-LKD) as Prime Minister. In response, the President called for an increase in executive powers to offset the new political balance of power. As a testament to the strength in the formal rules of the game, his efforts failed to generate any support from any party in Parliament in part because the previous LDDP government had tried a similar move in 1994; it proposed an amendment to decrease the number of seats in Parliament from 141 to 79 as a way to increase the percentage of seats in their favor. This effort failed as most other parties banded together to reject the amendment. The majority of voters followed suite when the government tried to initiate a referendum. These two failed challenges demonstrated the structural resilience in the Lithuanian political rules of the game.

Bureaucratic responsibilities also remained largely stable under such conditions, and VATESI enjoyed the benefits of autonomy and sufficient resources as a result. Unfortunately, Lithuania's state utility managing Ignalina struggled to collect electricity bills until 2001, and the limited profits jeopardized the pace of safety work. Since nearly fifty percent of Ignalina's production went into exports because of low domestic demand, and its largest consumers failed to pay their bills on time, plant management could neither invest in safety work nor pay Ignalina's staff their full salaries. Neighboring governments pressured to keep the plant running, no matter the safety condition, as they lobbied for its continued operation to provide cheap energy for their struggling economies.

Resistance to early closure had begun since the initial EBRD Grant Agreement in 1994. At that time, the Swedish and Finnish state electricity utilities had given Western legitimacy to those voices by stating that early closure would lead to serious economic costs to the entire region. Two years later, the Lithuanian Energy Institute (LEI) argued that "untimely closure of the Ignalina power plant would cripple the Lithuanian economy."[14] The LEI wanted Ignalina to run at least until 2020. Plant management agreed, stating that the plant could safely run for years and that the cost to shut down was too high.[15] These supporters gained some sympathy from an avid opponent to continued operation, Jan Nistad of the Swedish Nuclear Power Inspectorate. At the 1998 Stockholm conference, he expressed sympathy for the Lithuanians' plight saying that "It is easy for the bank to sit in London and negotiate, but if you don't have enough heat and your people are freezing, that's not so easy."[16] His comments perpetuated the resistance to early closure.

Widespread political opposition to the Grant Agreement's terms grew as the deadline approached. The Lithuanian Energy Institute and plant management voiced the loudest criticisms, but by 1997 parliamentary officials began to join the challenge to close Ignalina. In April, even as Prime Minister Vagnorius reiterated his country's commitment to honor the EBRD agreement, rumblings in the Seimas and the cabinet started shortly thereafter. LEI director Jurgis Vilemas acknowledged what appeared to be a two-faced policy toward Ignalina saying that "the Lithuanian government [was] discussing the possibility of keeping both of Ignalina's RBMKs running for another 15 years – in the corridors, yes; officially, no."[17] Two months later, Viktoras Valentukevicius, Vice Minister of Energy Affairs in the Economy Ministry, made the first official statement questioning the timetable for closing Ignalina, arguing that "when we become stronger, it will be feasible. Now, it is too big a burden to bear. If Lithuania is forced to shut down Ignalina as agreed, it would be catastrophic for the country. It's important to continue operation at Ignalina. Any radical decisions might be harmful."[18] He cited rising energy demands and recent safety assurances from plant management that the fuel channels could allow unit 1 to operate until 2005.[19] In addition, while the EBRD agreement stipulated that alternative power must be available to replace Ignalina before final closure, Lithuania's generating capacity from fossil fuels remained at the mercy of relations with Russia, the main supplier of oil and gas to the Baltic region. General reluctance to hasten the country's dependence on that uncertain source of energy further increased the short-term costs of compliance.

Despite Valentukevicius's statement, he did not represent official government policy. A few weeks later, Economics Minister Vytautas Bielauskas reassured an assembly of the Nuclear Safety Account that Lithuania would honor its commitments by completing all safety modifications and upgrades, following VATESI's recommendations and directives, and decommissioning Ignalina as agreed.[20] In response, LEI Director Vilemas continued to press his attack on the feasibility of early closure in March 1998. Vilemas vigorously decried the comments of EU Environment Commissioner Ritt Bjerregaard who conditioned EU membership on early closure. Vilemas blasted such conditions as political maneuvering without scientific basis. He argued that Lithuania had been young and inexperienced when it signed the Grant Agreement, and that with the pressure from Russia, "the most important thing was how to survive. We thought OK, the bank will help us study the reactors and if they are not all right, we will shut them down. But if there's new information, why not change the agreement?"[21] Unfortunately for Vilemas, testing at unit 1 the next month revealed severe cracking in some of the reactor coolant pipes. This major safety concern threatened the safety of the entire plant and Vilemas backtracked because of the severity of the problem.

The problem goes political

Opponents to Ignalina's early closure hoped to gain support from the new President, Valdas Adamkus. A former US Environmental Protection Agency official who returned to his native country to run for the presidency, Adamkus narrowly defeated LDDP candidate Arturas Paulauskas in the 1998 election. Like Brazauskas before him, the new President stressed the costs of closing Ignalina, but Adamkus seemed to be more sensitive to popular opinion, and as disillusionment towards the European Union grew after his election, Adamkus announced that "there [was] no condition that we have to shut down. There [was] only a request."[22] Sensing the dangers to the country's long-term goal of EU membership from such a statement, Homeland Union Prime Minister Vagnorius moderated the statement by stressing the value of accession even with closing Ignalina. Vagnorius' commitment resulted from the clear lack of alternatives to the European Union.

In contrast to Armenia where elites welcomed Russian influence, Lithuanians lamented their vulnerabilities to Russian pressure, recognized their need for Western export markets and assistance, and maintained a long-standing drive to rejoin Europe. When it became clear in November 1998 that the European Union would not budge on the 2004 date as it threatened to pass over Lithuania for membership if the government did not resolve the Ignalina issue quickly, some Lithuanian elites expressed doubts that it would happen, but the government complied nonetheless. EU membership would meet Lithuania's long-term needs even if it really hurt in the process. Stable rules of the game enabled elites to pursue those long-term benefits instead of focusing exclusively on their costs, some of which included angering the electorate in upcoming elections. They chose future rewards over immediate hardships because they could reliably count on a stable future for the political system, and their own personal involvement regardless of temporary election defeats.[23]

In that regard, President Adamkus could have backed out of Lithuania's commitment when his approval ratings soared to seventy to eighty percent during his first term in office since most Lithuanians opposed closing Ignalina.[24] Instead, he held to the dates and unlike Vladimir Meciar in Slovakia (Chapter 6), did not use that mandate to try to change the configuration of political power in his favor.[25] Whether his reasons stemmed from close policy alignment with Homeland Union, the ruling party in Parliament after 1996, or some other purpose, the rules remained unchallenged and unchanged. Normal changes in government took place in the context of stable formal rules of the game, which effectively constrained elites as they sought to advance their political agendas and personal ambitions.

In addition, stable informal rules established the procedures by which elites met those interests, specifically when internal party conflicts arose. In 1998, Prime Minister Gediminas Vagnorius and President Adamkus worked together to marginalize Vytautas Landsbergis, the Speaker of Parliament and a rival to the Prime Minister for leadership of Homeland Union. They did this by removing ministerial appointees and staff loyal to the Speaker, most importantly the Minister of Internal Affairs. Adamkus also removed another Landsbergis supporter, the unpopular Minster of European Affairs Laima Andrikiene. Their actions followed established legal procedure as part of a larger contest for control over foreign policy between the President and the Speaker. Prime Minister Vagnorius had a central role in the contest given Homeland Union's coalition majority; the Prime Minister had almost total governing power as long as he controlled the party machinery. Vagnorius could effectively determine both domestic and foreign policy with Landsbergis out of the way, a clear benefit to the President.

However, even the Prime Minister was subject to the informal rules of the game. In 1999 President Adamkus used his enormous popular mandate to force Prime Minister Vagnorius to resign. The President's criticisms stemmed from Vagnorius's continued support of Economics Minister Vincas Babilius, whose privatization measures caused growing public opposition to the government. In an April 19 televised address, Adamkus said he no longer trusted Vagnorius, "claiming that he could not adequately perform his presidential duties while Vagnorius held the premiership."[26] He did not use his constitutional power to ask the Seimas to remove him; just that Vagnorius do the right thing and resign for the good of the country. The Prime Minister countered that Adamkus was violating the separation of powers clause in the constitution, even though the President was within his legal authority to do so. He simply took advantage of his informal power to force Vagnorius out of office.

The Prime Minister still had all the formal power he needed as long the ruling coalition continued to support him. He led the government and the government controlled Parliament. This was evidenced two days after Adamkus's speech when Homeland Union and its coalition partner, the Lithuanian Christian Democrats (LKD), voted their support for the Prime Minister in a non-binding resolution. Unfortunately for him, the people kept their views of Vagnorius and his popular support dwindled before the 2000 parliamentary election. The outcome was not hard to predict with an unpopular Prime Minister and a President voicing the complaints of eighty percent of the voters. TS and LKD followed the prevailing political winds and Vagnorius resigned a week later. Most critically for

Ignalina, despite the political gaming, the rules remained stable and governance procedures continued as before.

After Vagnorius resigned, the President appointed Rolandas Paksas as his replacement, although he too fell victim to unfavorable economic conditions and resigned shortly thereafter. Andrius Kubilius took over for the last year of Homeland Union's term in office, and managed to steer the economy more gently than had either Vagnorius or Paksas before him. Yet the rapid turnover in the Prime Minister's office had not evidenced problems in the mechanisms by which the changes occurred, which remained predictable and long-standing.

The success of the political system did not translate to success for Homeland Union though. The revolving door in the Prime Minister's office severely damaged the party's reputation, and it became increasingly unpopular with the electorate. The party had not been particularly popular for some time. Its earlier election victory in 1996 was "due mainly to disillusionment with the government rather than strong enthusiasm for the opposition parties."[27] LDDP's "failure in office to honor many of its electoral promises undermined its socialist credentials with the electorate in 1996. Worse was the widespread belief in the corruption of its leading politicians."[28] Nonetheless, corruption did not produce the same instability in the informal rules of the game as seen in Bulgaria (Chapter 5). Since Lithuania's independence, ruling political parties have faced protest voting but have always been able to establish authoritative governments under the procedures established by the durable party system. The 2000 Liberal Union of Lithuania (LLS) coalition government was the only government to fail in that capacity, but did so as a minority coalition, and with a plurality of votes the Social Democrats took over shortly thereafter. The predictable transfer of power testifies to the stability of the rules of the game and the long-term thinking they produced.

In that context, when Brazauskas and his LDDP reclaimed power in June 2001, the new Prime Minister maintained the same EU-centered policies fostered by the Liberal Union. Six months earlier he had signed a joint government-opposition statement supporting speedier integration into the European Union in large part because the economy saw real improvement, and discontent with EU austerity measures had begun to diminish, although they were still unpopular.[29] In addition, vulnerability to Russian pressure and the need for Western assistance to strengthen the nascent democratic regime made EU accession a consistent priority for each successive government. Yet, the Lithuanians still tried to change the minds of their counterparts in the Nuclear Safety Account with arguments from across the party spectrum that Ignalina should be allowed to continue operations.

Ignalina's last gasp

At a celebration of the plant's first fifteen years of operation, Prime Minister Vagnorius told a conference in Vilnius that he still did not grasp the necessity of closing the plant before the end of its design lifetime, calling into question the need to "sacrifice the economy of Lithuania for political motives."[30] Accordingly, as late as October 1999 the Seimas passed a new energy bill that called for closure of unit 1 in 2005, outside the Grant Agreement's deadline, and conditioned the decommissioning decision for unit 2 as part of the regularly scheduled review of the country's energy policies in 2004. European Commission officials applauded the resolution on unit 1 in principle, but still had harsh words about missing the 2004 deadline and delaying the decision on unit 2.

Prime Minister Paksas countered and followed his predecessor's course by asserting the right to determine national energy policy, something clearly articulated in both the Euratom charter and the Convention on Nuclear Safety. In April 2001, he stated that "Lithuania [would] not be pushed by the EU into setting a date for the shutdown of Ignalina-2 before 2004. We all remember the Chernobyl accident. Ignalina-2 is different. It is absolutely safe. It's absolutely not comparable to Chernobyl."[31] In response to these challenges, the European Commission made it clear that compliance with international agreements mattered more than different definitions of safety, reminding the Lithuanians of the cost of missing EU accession deadlines.[32] Aware of those costs and the country's need for the long-term benefits of compliance, Paksas stopped dissenting and told *Nucleonics Week* in a June 2001 interview that he saw clear "indications from the European Union that this matter has to be speeded up."[33] At a June 15–16 European Council summit meeting in Sweden, Paksas further moderated his earlier position, stating that he "simply quoted the law. However, we always have to balance between our laws and the EU requirements."[34] His reversal signaled a growing acceptance of Ignalina's inevitable closure, and although incoming Prime Minister Brazauskas also questioned the path of high compliance, he too stayed on it.

Shortly after taking power in July 2001, Brazauskas told a European Commission summit meeting that the new government could not make a decision on the costs of closing unit 2 at that time. He had not contradicted the previous government's acceptance of early closure, but neither had he emphatically endorsed the EU's preferred date. The Prime Minister later repeated earlier claims that closing unit 2 so close to the shutdown of unit 1 would cripple the economy.[35] However, others recognized the need to give in to the European Union's demands even if Brazauskas resisted. Speaker of Parliament Arturas Paulauskas, who formerly opposed Brazauskas's LDDP party but later joined it to form the presiding government coalition, stated clearly that " 'Lithuania has to set a date for shutdown of the second unit if it wants to keep its EU membership negotiations on track.' But he noted that Lithuanian nuclear experts say the unit can operate safely until 2012."[36] As with all previous reluctance to comply, Brazauskas's government relented by the end of the year, and the Seimas asked the government to prepare legislation to change the national energy law enabling a decision on unit 2 to be made before 2004, thereby cutting its operating lifetime well short of the 2012 mark.

Stable rules of the game allowed elites from different parties to see accurately the costs and benefits awaiting them, the most important of which had always been EU membership. However, the EC overseeing accession negotiations recognized the need for some carrot to balance the stick, so in a last-minute concession in mid-1999 stated that it would begin accession negotiations before unit 2 had completely shut down, as long as it happened by 2009.[37] The revision emboldened some future attempts by Vagnorius, Paksas, and Brazauskas to wiggle out of the date, but none went so far as to threaten compliance. They understandably blustered, but still kept on track with decommissioning and high compliance.

In keeping with the previous half-hearted attempts, President Adamkus stated in a February 2002 public address that he did not approve of the government setting specific closing dates for closing Ignalina unit 2. He argued instead for "a rational look at [Lithuania's] energy policy, assessment of all possibilities and a decision based on this. I do not accept pressure at all. The government must be resolute."[38] His position was not surprising given the upcoming presidential elections one month away, and the fact that EU membership persisted as a contentious issue among the electorate.[39] The impact of Adamkus's resistance diminished quickly when he moderated his position a few days later at a London meeting with EBRD and EC officials, although he still made one final attempt to change the terms of Lithuania's international commitments. Like the previous attempts, it lasted for only a brief time and then passed away. As per the West's demands, on June 11, 2002, the government of Lithuania agreed to close Ignalina-2 by 2009 and thereby close the energy chapter of the EU accession agreement. Four months later, the Seimas approved a new energy bill confirming the same timeline, and the EBRD established a new decommissioning fund. Lithuania joined the European Union in 2004, closed unit 1 the same year, and unit 2 followed five years later as ordered. As a result, Lithuania had fully complied with its international agreements.

Conclusion

High compliance in Lithuania did not result from safety culture because it remained consistently opposed to early closure. Regulatory Chairman Kutas indicated to me in September 2002 that he saw no reason the plant could not operate for many years beyond the EU demands if membership dropped out of the picture and safety became the only evaluative tool for determining Ignalina's fate.[40] The thought of two 1,500-megawatt RBMK reactors operating twenty or more years without a containment structure rightly raised concerns, but it also indicated the kind of risk acceptant safety culture that persisted among Lithuania's nuclear and political officials. Each year that operations continued would be one more role of the dice, and Kutas implied the Lithuanians would willingly accept that chance if they did not value EU membership so highly. Like its Baltic neighbors, Lithuania needed the European Union for economic development and political legitimacy in the face of historic Russian pressure, and went against its safety culture to get it.

The EC and NSA donors responded by giving nearly $500 million to support the decommissioning of Ignalina's first unit once the Lithuanians made the decision, with the promise that more would be available for unit 2 in the future; Euratom loan ceilings increased to $6 billion to cover any remaining costs that might arise over the next decade. The 2002 EU Regular Report stated that "the European Union, for its part, recognizes that the decommissioning of the Ignalina Nuclear Power Plant will have to continue beyond the current financial perspectives and that this effort represents for Lithuania an exceptional financial burden not commensurate with the size and economic strength of the country."[41] The report continued that the European Union would offer assistance as needed to finish the job after Lithuania's accession. As a result, Foreign Minister Antanas Valonius summed up his government's response to the offer: "Lithuania got the deal it wanted on Ignalina."[42] The country was officially invited to join the European Union in December 2002 as a reflection of Lithuania's progress and commitment to high compliance.

The stability of the formal and informal rules enabled Lithuanian elites to make the hard decision to close Ignalina even though they considered it a safe and vital state resource. This achievement helped set Lithuania apart from the other Eastern European recipients of international nuclear safety assistance, as did the stability of the rules of the game. Table 4.1 indicates the levels of rules stability and compliance accordingly.

Table 4.1 Lithuania Independence (1991)–Decommissioning (2009)

Rules Stability	*Clarity*	*Predictability*
Formal	Yes	Yes
Informal	Yes	Yes
Conclusion: Formal rules experienced no serious threats to the constitutional delegation of power or transfer mechanisms. Informal rules operated based on predictable party positions, regular elections producing reliable alternations in power and clear bureaucratic responsibilities for Ignalina. Lithuania indicates **HIGH STABILITY**.		
Compliance	• Completed safety upgrades • Sufficient legal autonomy and funding for VATESI, increasing personnel, and operation within parameters of CNS • Full compliance with Grant Agreement – unit 1 closed on time in 2004, unit 2 in 2009	
Conclusion: Lithuania indicates **HIGH COMPLIANCE**.		

Several factors contributed to the stability of the rules from the beginning of the post-communist period. First, the prospect of EU membership incentivized democratization and political stability, and Lithuania's political elites recognized early on the necessity of hard choices to break out of Russia's orbit, even with some understandable vacillations pulling them away from the West at times. That common recognition enabled the rules of the game to become quickly entrenched and widely accepted. Second, pro-Western policy convergence between the major parties also helped mitigate the unpredictable and sometimes violent policy shifts found in other Eastern European countries. Third, the country's historic legacy of democratic governance prior to World War II, and an almost universal sense of nationalism driving the country's demands to secede from the Soviet Union, helped clarify the costs and benefits of compliance despite the general reluctance to do so fully. As a result, Lithuanian compliance had many things going in its favor, all of which made the costly decision to close Ignalina possible in the first place. More than any other country in the region, while the Lithuanians had the most to lose, they also had the most to gain, and the stable political system allowed them to see the choice clearly. In the end, they remained reluctant, but still obeyed.

Notes

1 *Nucleonics Week*, May 20, 1993: 3.
2 Krickus (1997): 294 gives more detail on the negotiations during the independence movement.
3 Most of Lithuania's political parties came out of Sajudis and/or the Communist Party.
4 Girnius (1999): 51. There have been minor violations such as appointing generals as ministers of interior and allowing Kazys Bobelis (a US citizen) to hold office before he renounced his US citizenship (Ibid., 52). Both violated the constitution but did not affect the stability of the formal rules of the game.
5 In addition, Ignalina plant management and VATESI were at the mercy of RDIPE (the Russian RBMK design institute) to send design schematics and information necessary to integrate new safety modifications with existing systems. Povilas Vaishnis, the first director of VATESI through the early years of assistance, lamented that the Russians would often delay requests for information or send incomplete data (*Nucleonics Week*, September 8, 1994: 15) Interestingly, most of the personnel at Ignalina were Russian, or at least they were described as Russian-speakers who did not speak Lithuanian.
6 If there was a way to construct a cheap containment structure and change the reactor core from graphite moderated to some other form, like sodium moderated fast breeder or pressurized water, then it might have been possible for the plant to operate until the end of its design lifetime. Neither of those options proved financially feasible or technically certified, making early closure the only acceptable option for Ignalina according to the West.
7 *Nucleonics Week*, September 11, 1997: 12.
8 *Nucleonics Week*, September 25, 1997: 13.
9 See WENRA (2000): 59–70 for a thorough assessment of VATESI's status. An interesting challenge to that status came early in November 1995 when President Brazauskas stated that Ignalina would not shut down before 2000, but the decision was supposed to be based on VATESI's assessment regardless of political foot-dragging. However, even though the President made his statement, the decision still rested with VATESI. Plant personnel, political elites, and Western officials waited on VATESI, and although some attempts by the European Commission to speed up the process were offered, they were rejected out of hand. VATESI Chairman Kutas told me his organization was going to work independently if necessary, at its own pace to make sure everything was done accurately and in compliance with their mandate, and they did (Saulius Kutas, in discussion with the author, September 2002). Carol Kessler also had high praise for Kutas and VATESI saying they had done a tremendous job, especially with limited resources (Carol Kessler, in discussion with author, July 2002).
10 In an interview with *Nucleonics Week*, VATESI Chairman Vaishnis responded that his organization was completely up to the task. "We didn't deserve such an evaluation. We started practically at zero and worked very hard in the past five years to develop VATESI. There wasn't a basis for some of these conclusions" (*Nucleonics Week*, March 13, 1997: 6). Vaishnis resigned in protest later that year.
11 Gelazis (2001): 166.
12 "By November 1995 the Lithuanian press had reported 43 corruption scandals in the government." Combined with the "serious decline in living standards of large sections of the population" this led to LDDP's downfall, despite improvements in general economic indices (Lane, 2001: 143).
13 Ibid., 143–144. Slezevicius had also received a much higher rate of return on his deposits, another insider benefit that further fueled the ire of an already irate public (Krickus, 1997:311).
14 *Nucleonics Week*, July 16, 1998: 7.
15 See *Nucleonics Week*, November 28, 1996: 4 and *Nucleonics Week*, January 13, 2000: 11 for two examples.

16 *Nucleonics Week*, August 22, 1996: 6.
17 *Nucleonics Week*, May 15, 1997: 11.
18 *Nucleonics Week*, June 5, 1997: 6.
19 Ultimately, VATESI called the shots, but while the regulator worked through the mountain of technical requirements for the SAR, Shevaldin conducted tests of his own.
20 Carol Kessler, in discussion with author, July 2002; *Nucleonics Week*, June 19, 1997: 13.
21 *Nucleonics Week*, March 19, 1998: 5.
22 *Nucleonics Week*, June 4, 1998: 9.
23 It was a good thing they could because the short-term costs for compliance stood at nearly $4 billion dollars, not including the further long-term costs from the loss of electricity export revenue and the need to buy more fossil fuels.
24 Clark and Adams (1998): 126. Unlike Brazauskas or Paulauskas before him who tried to change the formal rules of the game, Adamkus only tried to consolidate ministerial portfolios by reducing the number from seventeen to eleven (the Seimas agreed to reduce the number to fourteen). The change to the informal delegation of power was accepted by all parties as necessary and a compromise reached within the legal means proscribed by the constitution.
25 His actions and the consequences matter more than the reasons he chose them. In contrast, Paulauskas made it clear that he intended to change the constitution to strengthen the President's powers at the expense of the government. "This proposal alone was significant and would have substantially reduced Parliament's role in Lithuanian politics." (*East European Constitutional Review*, 1998). Paulauskas also wanted to reinterpret Article 84 to enable him to dismiss the current government and appoint a new one. The Constitutional Court disagreed, ruling that only after a parliamentary election when the power that invested the government changes, does the government also need to resign. After a presidential contest, the President simply confirms Parliament's approval with a vote of confidence. Accordingly, the mechanisms of power transfer and delegation of power between branches of government remained predictable and unchallenged. Paulauskas was more of a political rogue than representative of a widespread movement to challenge the rules of the game. His election defeat maintained stability in the system, but even if he had won it is unlikely his efforts would have succeeded. The rules had become solidly entrenched as the failure of earlier attempts shows.
26 Ibid.
27 Lane (2001): 146.
28 Ibid., 134.
29 Karatnycky (2002): 255. Representatives were from LDDP-turned-Social Democrats; centrist parties Homeland Union, Center Union, and Lithuanian Christian Democrats; and the "free-market" Lithuanian Liberal Union and Lithuanian Democratic Party.
30 *Nucleonics Week*, April 15, 1999: 8.
31 *Nucleonics Week*, April 5, 2001: 3.
32 Vagnorius had asked for a new international review panel to see if the plant was safe and could operate beyond the closure timetable. European Commissioner Hans van den Broek sent a letter back saying no panel was needed, only Lithuanian compliance with the NSA agreement that plant management not rechannel the fuel rods – a position reiterated at the October 7 NSA donors meeting.
33 *Nucleonics Week*, June 21, 2001: 1.
34 Ibid.
35 *Nucleonics Week*, November 15, 2001: 16.
36 *Nucleonics Week*, September 27, 2001: 4.
37 The original EBRD agreement stated a time period of 1998–2004 for unit 1. Unit 2 was to follow shortly thereafter assuming a similar rate of gap closure for its fuel channels. When VATESI determined that the channels could last longer if the plant continued

to operate below full power – which it did until late 2001 due to weak demand and limited payment from the plant's customers – until 2005 and 2009 respectively, the dates were pushed back.

38 *Nucleonics Week*, February 28, 2002: 9.

39 "The director of the Vilmorus opinion research center, Vladas Gaidys, told RFE/RL that euroskepticism – roughly half the population supports EU membership compared to much higher levels in the early and mid 1990s – is on the rise in Lithuania because of the unexpectedly small agricultural subsidies and the Ignalina issue. Although people are worried about the plant's safety risks, they are also concerned about the economic consequences of shutting down the plant altogether." (*Radio Free Europe/ Radio Liberty*, 2002.) Also, LDDP had won the previous parliamentary election in 2000 on a less EU-friendly policy. Brauzaskas later saw the writing on the wall and changed positions.

40 Saulius Kutas, in discussion with the author, September 2002.

41 European Commission (2002): 88.

42 *Nucleonics Week*, December 12, 2002: 14.

References

Clark, Terry and Andrea Adams. "The Year of the President." In *Holding the Course: Annual Survey of Eastern Europe and the Former Soviet Union*. edited by Peter Rutland, Chapter 27. London: M.E. Sharpe, 1998.

East European Constitutional Review. "Constitution Watch: Lithuania." 7, no. 1 (1998).

Economist. "Regular Report on Lithuania's Progress towards Accession." October 2002.

Gelazis, Nida. "Institutional Engineering in Lithuania: Stability through Compromise." In *Democratic Consolidation in Eastern Europe: Institutional Engineering, Volume 1*, edited by Jan Zielonka, chapter 6. New York: Oxford University Press, 2001.

Girnius, Kestutis. "Democracy in Lithuania." In *Democratization in Central and Eastern Europe*, edited by Mary Kaldor and Ivan Vejvoda, chapter 4. New York: Pinter, 1999.

Karatnycky, Adrian, Alexander Motyl and Amanda Schnetzer, eds. *Nations in Transit 2001: Civil Society, Democracy and Markets in East Central Europe and the Newly Independent States*. New Brunswick: Transaction Publications, 2002.

Krickus, Richard. "Democratization in Lithuania." In *The Consolidation of Democracy in East-Central Europe*, edited by Karen Dawisha and Bruce Parrott, chapter 8. New York: Cambridge University Press, 1997.

Lane, Thomas. *Lithuania Stepping Westward*. London: Routledge, 2001.

Nucleonics Week. May 20, 1993: 3; September 8, 1994: 15; August 22, 1996: 6; November 28, 1996: 4; March 13, 1997: 6; May 15, 1997: 11; June 5, 1997: 6; June 19, 1997: 13; September 11, 1997: 12; September 25, 1997: 13; March 19, 1998: 5; June 4, 1998: 9; July 16, 1998: 7; April 15, 1999: 8; January 13, 2000: 11; April 5, 2001: 3; June 21, 2001: 1; September 27, 2001: 4; November 15, 2001: 16; February 28, 2002: 9; December 12, 2002: 14.

Radio Free Europe/Radio Liberty "Lithuania: EU Urges Closure of Ignalina Nuclear Plant." April 4, 2002.

Western European Nuclear Regulators Association. "Nuclear Safety in EU Candidate Countries." October 2000.

Additional Reading

BBC News. "Adamkus Leads Lithuania Poll." December 23, 2002.

——. "Lithuania Votes for President." January 5, 2003.

——. "Shock Result in Lithuania Poll." January 5, 2003.

Clemens, Walter. "The Baltic Reborn: Challenges of Transition." *Demokratizatsiya* 6, no. 4 (Fall 1998): 710–711.

International Atomic Energy Agency (IAEA). Convention on Nuclear Safety, 1994.

Economist. "Throw Them Out: Lithuania," October 12 1996.

——. "Algirdas Brazauskas; Charlemagne." August 25 2001.

European Commission. "Regular Report on Lithuania's Progress towards Accession." 1998

——. "Regular Report on Lithuania's Progress towards Accession." October 1999.

——. "Regular Report on Lithuania's Progress towards Accession." November 2000.

——. "Regular Report on Lithuania's Progress towards Accession." November 2001.

Framework Agreement between the Republic of Lithuania and the European Bank for Reconstruction and Development Relating to the Activities of the Ignalina International Decommissioning Support Fund in Lithuania. 2001.

Grant Agreement between the Republic of Lithuania and the European Bank for Reconstruction and Development, Ignalina 1 and 2 Nuclear Safety Project. 1994.

Marsh, Christopher. "Power Capabilities, External Recognition and Sovereignty: The Case of Lithuanian Independence." *East European Quarterly* 35, no. 1 (March 2001): 109–127.

Pettai, Vello and Marcus Kreuzer. "Party Politics in the Baltic States: Social Bases and Institutional Context." *East European Politics and Society* 13, no. 1 (Winter 1999): 148–152.

5 Bulgaria, what a strange trip it has been

"In Bulgaria the political situation changed every year. The political instability affected the industry because it changed personnel each time. This was very detrimental to the nuclear industry."

—Yanko Yanev, former Chairman, Bulgarian State Committee for the Utilization of Atomic Energy for Peaceful Purposes (CUAEPP), September 2002

Introduction

In 1993 the Bulgarian government signed an agreement with the EBRD to close the four oldest VVER 440-230 reactors at the Kozloduy nuclear power plant within five years. The agreement required replacement power from two next-generation VVER 1000 reactors awaiting completion and the modernization of several fossil fuel plants. However, neither alternative was ready by 1998 and Kozloduy continued to operate past the deadline. Blame did not rest on Western donors though since Bulgarian elites across the political spectrum repeatedly delayed implementing the costly requirements necessary for compliance, specifically failing to raise electricity prices and invest available money to replace their unsafe nuclear reactors.

After missing the first deadline, the Bulgarian government again committed itself to close units 1 and 2 by 2003, while leaving the fate of 3 and 4 to be decided afterward. Soon thereafter, a Memorandum of Understanding (MoU) signed with the European Commission in 1999 contained a clear promise that the shutdown date for units 3 and 4 would take place no later than 2006, several years earlier than the reactors' thirty-year design lifetimes. The MoU referenced Bulgaria's prior commitment to the EBRD and indicated that the EC would not accept a repeat performance. Units 1 and 2 closed on time in 2002, but despite their commitment to close the third and fourth units in 2006, the Bulgarians tried repeatedly to wiggle their way out of the binding agreement. More problematically, some government officials even questioned the need to comply with the shutdown order for 1 and 2.

This kind of compliance behavior corresponded to the rhythms of the Bulgarian political system. Janus-faced throughout its first post-independence decade, one side of Bulgaria's system stood as "an island of stability in the Balkans,"[1] due to the strength of formal democratic institutions established during the inclusive political negotiations of 1990. Popular acceptance of a democratic system and a lack of politically motivated violence reinforced those strengths.[2] Unfortunately, the system's other face revealed the day-to-day instability of a country seriously disrupted by the contest between the Union of Democratic Forces (SDS) and the Bulgarian Socialist Party (BSP). The competition between the two took place according to the formal provisions laid out in the constitution, but the ways the rivalry played itself out caused a great deal of uncertainty and instability in the informal rules of the game. Even the introduction of a third player, former King Simeon II, to the political process failed to stabilize the informal rules. As a result, the Bulgarian political system failed to produce authoritative governments of any significance throughout the period of nuclear safety compliance, often making governance erratic and unpredictable.

In contrast, the constitution clearly defined and established the formal rules of the game with clear differentiations of delegated authority between the President and Parliament, which Bulgarian elites did not seriously challenge; the few isolated attempts proved ineffectual. The constitution also allowed for pre-term elections and votes of no-confidence, procedures that in themselves did not introduce instability. However, while the structures of government worked well, informal governance suffered due to unreliable procedures for assigning ministerial appointments, unclear bureaucratic responsibilities between government ministries, and unpredictable policy reversals between successive governments.[3] Most elites could not count on regular tenures in office, whether short or full-term, nor could they reliably predict whether incoming governments would discard or sustain their policy programs after election victories. This type of uncertainty did not force elites into short-term thinking, but neither did it lead to long time horizons either as there was no obvious way to hope for a brighter, more stable future. The condition of medium rules stability, with stable formal rules and unstable informal rules, produced medium levels of compliance. As a result, the Bulgarians completed safety upgrades and established regulatory sufficiency on time, but failed to meet several conditions for high compliance.

Janus appears

The 2001 parliamentary elections marked the completion of the first full term in office for any post-communist Bulgarian government. The preceding four governments had all been forced out in early elections or votes of no-confidence with the Democrats and Socialists trading places each time. Only after former King Simeon's National Movement for Simeon the Second (NDSV) pulled off an election upset did the country finally get a new ruling party. Simeon's luster faded quickly though, and the decade-old governance problems continued.

In contrast, the country had earlier left behind its former single-party rule with a relatively peaceful transition at the beginning of the post-Soviet era, one characterized by elite cooperation between reconstructed communists and democratic reformers. The "round-table" talks produced a solid constitutional basis for new governing institutions with clear divisions of power between Parliament and the President, predictable mechanisms of power transfer for those institutions, and a lack of challenges to the established delegation of institutional authority. The first elections for the new Grand National Assembly enabled BSP to defeat the loosely coordinated Democratic groups, although less than a year later SDS took control of Parliament with a slight four-seat advantage over the Socialist Party. Early the next year SDS candidate Zhelyu Zhelev defeated the Socialist candidate for the presidency, Velko Valkanov, by a slim majority (fifty-three percent) in the second round of voting, seemingly indicating a clear path ahead for Bulgarian policy. Unfortunately, the turn-around vote for SDS in the parliamentary and presidential elections less than two years after BSP won control of the legislature presaged troubling events to come. Whereas both sides had worked cooperatively to end communist rule, deep and abiding polarization within the political elite had begun to surface, revealing rivalries that the initial independence movement had merely masked.

While the Socialists lost some of their early support, the closeness of the election results in 1991 and 1992 reflected a division between Western-leaning free marketers, and those who valued close traditional ties with Russia and the maintenance of state-run economic programs. Policy differences represented degrees rather than polar opposites though. Neither party advocated extreme isolation or complete integration with the West, instead preferring to keep one foot in and the other out. Yet despite this common pragmatism toward international orientations, SDS and BSP offered diametrically opposed domestic policies. Critically for the country's nuclear safety compliance, neither party always changed their opponents' policies once in office. Instead, piecemeal, episodic changes occurred irregularly so elites could neither count on short-lived policies, nor could they reliably plan long-term policy goals given the potential of electoral defeat and a potential policy reversal. Therefore, while one party expressed doubts about the benefits of Western alignment, both groups faced a general uncertainty in the informal rules, thereby diminishing their willingness to endure the highest costs of compliance to gain the long-term benefits of high compliance.

Antipathy between the Democrats and Socialists grew in that context, and the first SDS government could only govern from a minority position after joining other anti-BSP parties for support. Aiding those efforts, the Socialists initially isolated themselves to such a degree that they could not overcome the four-seat deficit to regain control of the government. However, SDS did not have much time to consolidate its control as internal divisions began to pull the coalition apart. Several members defected and joined a vote of no-confidence sponsored by the DPS (Movement for Rights and Freedoms) on October 28, 1992, and Filip Dimitrov's SDS government collapsed after losing the vote. Internal party maneuvering within SDS and between the two main parties and DPS, which had become increasingly important as a coalition swing-vote, helped to produce widespread uncertainty as to who would rule and how long they would last.

The constitutional provisions for power transfer offered a temporary solution to the worsening crisis of governance. Despite assigning very limited powers to the presidency, notably the ability to dismiss Parliament if it fails to elect a new government three times and thereby triggering early elections, the President can also call for the formation of a new government. SDS President Zhelev used that power and announced that his party could try again in late 1992. Yet even with strong presidential backing, the party leadership could not win enough support for Dimitrov to regain control of the government. DPS responded and put forward a compromise candidate for Prime Minister, Liuben Berov, who convinced several SDS and BSP members to support his selection in 1993. However, like the SDS government before him, Berov's team proved equally unable to resolve the country's underlying economic and social problems, prompting his resignation two years later. The configuration of power in the legislature remained the same though as neither main party succeeded in convincing smaller parties to join them in a new government. As a result, President Zhelev threatened to dismiss the Parliament in the midst of the crisis, arguing that if the legislature could not govern, the country might be better off replacing the existing parliamentary republic with a presidential one.[4] His proposal did not represent a serious challenge to the constitutional order since he did not call for public referenda or make good his threat to dissolve Parliament, and judicial review rejected the plan in any account.

Even with this rare exception, no serious changes to the formal rules ever occurred despite the growing crisis in governance. Yet it became evident after Berov's resignation that the constitution had a key weakness, namely its lack of clear stipulations for how the government should function in the event the President dismissed it, a growing possibility as the morass in Parliament continued. This vagueness produced uncertainty as to how power would be transferred, and no one at the time knew for certain who would be in charge before the next elections produced a new government.[5] Surprisingly, in a rare show of unity, SDS and BSP agreed to early elections in December 1994 to avert that potential problem, although the Democrats may not have agreed had they known beforehand how extensively the electorate blamed them for the country's myriad problems.

Kozloduy: Bulgaria's boggart

Bulgaria's sufferings extended beyond political turmoil and economic upheaval. As early as June 1991 IAEA safety experts reported that despite their technical differences, the original four VVER Kozloduy reactors had similar problems as the fatally-flawed Chernobyl plant, and therefore could produce deadly accidents any time. Severe deficiencies in operational management, uncoordinated and inadequate staff training, woefully out-dated equipment, and insufficient safety measures that had been hastily constructed and poorly implemented all seriously threatened the safety of the Bulgarian facility. Most troubling was the lack of containment structures given the likelihood of an inevitable catastrophe.

One year later the G7 issued its edict that all first-generation Soviet reactors should be shut down before the end of their design lifetimes. In particular, Kozloduy received the lion's share of the G7's assistance package due to its poor conditions, and that was saying a lot given the others in the group. In particular, the West offered significant assistance for short-term safety upgrades as an interim step during the time it took to coordinate closure agreements. Upgrades were never intended to keep the plants running as the only "reasonable solution [was] definitive shutdown,"[6] according to German Minister for Environmental and Nuclear Safety, Klaus Toepfer, a lead contributor in European Commission discussions on Kozloduy.

However, the four 440-230s operating at Kozloduy did not share identical levels of safety. Units 3 and 4 had some additional measures built into the plant during their construction, but the first and second units embodied all that could go wrong with early Soviet-designed reactors. Aside from lacking containment structures, the greatest problem concerned embrittlement of the reactors' pressure vessels. Lack of sufficient maintenance over the years led Western safety experts to conclude that in the event of a primary system leak, the sudden inrush of cold water could cause the pipe to rupture, thereby threatening the safety of the entire cooling system. The emergency cooling system could fail altogether if the rupture was severe enough, leading to a meltdown in the fuel rods. Making matters worse, the fire protection systems remained completely inadequate with flammable wooden doors separating the generator rooms from command and control offices.[7] A fire at the plant in September 1992 greatly heightened international fears.

The 1992 SDS government initially approved limited safety work under the EU-Phare program, but the scale was relatively small compared to the EBRD's later proposal to the DPS coalition government in 1993. The Bank wanted Bulgaria to set the standard for later work as the first recipient of a Grant Agreement under the Nuclear Safety Account (NSA). The Agreement gave Bulgaria nearly a third of the $90 million initially donated to the Bank's fund, and in return the government accepted the terms of a strict closure agreement calling for early decommissioning of all four reactors by 1998. Early shutdown required available alternative energy supplies either through domestic production or imports, both of which required significant restructuring of the country's energy sector to improve efficiency and reduce demand to meet that condition. Bulgaria had one of the most inefficient energy markets at the time, something that persisted for more than a decade.[8]

As with later agreements, Western governments and safety experts argued that Bulgaria needed to do three things: 1) correct the immediate safety deficiencies at Kozloduy, 2) improve the autonomy and financial resources of the nuclear regulatory authority, and 3) take the necessary steps to close Kozloduy's most dangerous reactors (units 1–4) before the end of the millennium. However, despite their promises, Bulgaria's political and nuclear officials resisted shutting down any of Kozloduy's reactors, and vigorously opposed burdening the fragile economy with costly price increases to pay for early closure. Immediate concerns overshadowed the future benefits of compliance because, while stability in the formal rules of the game offered some certainty about the political situation, instability in the informal governance procedures made long-term thinking unlikely. As a result, safety improved quickly and the regulatory agency began to improve, but decommissioning remained a dubious outcome.

Central to all three levels of compliance, the nuclear regulatory agency initially functioned as a semi-autonomous Committee for the Utilization of Atomic Energy for Peaceful Purposes, beholden to the Energy Ministry for budget and oversight but still able to operate independently within the Ministry's general parameters.[9] The group's first post-independence chairman, Yanko Yanev, readily acknowledged the need for outside assistance given the plant's condition and the government's inability to rectify the problems. Yanev affirmed his commitment to operational safety and even temporarily shut down unit 4 in May 1993 to upgrade several systems, even though it was the safest reactor at Kozloduy. His agency attempted to make Kozloduy both safer and more profitable, but as with the two-sided political system, on the one hand Yanev gladly received Western money, while on the other vehemently resisted Western pressure and his own government's commitment to close any reactor at Kozloduy before the end of its design lifetime.[10] He also publically disagreed with the government's acceptance of the 1993 Grant Agreement requiring shutdown of Kozloduy's first two reactors, and worked at cross-purposes to ensure full-term operation as a critical way to get Bulgaria out of its economic woes.[11] Safety clearly mattered to Yanev and the nuclear regulators he oversaw, but Bulgaria needed Kozloduy's power generation and Yanev told me that he would run Kozloduy with or without safety upgrades, so great were the country's needs and the government's unpredictability at the time.[12] Even though Yanev only served as chairman for a few years, his example continued through successive governments and regulatory leaders, in part due that unpredictability and a desire to keep one thing certain – Kozloduy as Bulgaria's symbol of national pride.

Yanev's type of short-term thinking pervaded the political and nuclear elites. Nikita Shervashidze, Chairman of Bulgaria's Nuclear Energy Committee (NEC) that managed Kozloduy, argued that his country should not become a "Trojan Horse that [would] let Western banks impose unacceptable conditions on east European countries."[13] He adamantly resisted the EBRD's demand that units 1 and 2 close by 1996, to be followed by units 3 and 4 two years later, calling it "absurd to require the closure of units 3 and 4, more modern than the first pair, and added that closure of units 1 and 2 could be considered only after 1998."[14] Shervashidze did not agree to close them in 1998, only that the decision might be considered at that time. Therefore, despite a firm commitment to close 1 and 2 by 1998, the prospects of full Bulgarian compliance remained doubtful from the start.

One possible explanation for the Bulgarian resistance could be the perception of exit clauses in the Grant Agreement. The Bank had conditioned closure on several factors that allowed for some flexibility on the exact decommissioning date in 1998, which the Bulgarians may have interpreted as allowing for flexibility in general. However, the Agreement had a firm commitment to shutdown before the end of the design lifetime, and Western expectations of full compliance by the end of 1998 for units 1 and 2 never wavered. Francois Demarcq, NSA Administrator at the time, stated that the Bulgarians had "officially committed to following the path of the pact of trust,"[15] what he later called a "gentlemen's agreement."[16] Six months later France's nuclear regulator, Andre-Claude Lacoste said that "Bulgaria promised the EBRD to put in place investments in new capacity that would allow closure of Kozloduy 1 and 2 around 1998, in exchange for $28 million in grants for equipment purchases. But donors are now beginning to doubt the strength of the Bulgarian commitment."[17] That doubt grew as the Bulgarians took advantage of the situation whenever possible, willingly accepting technical equipment to modernize Kozloduy while refusing the harder decisions.

Western assistance had been offered to enable a limited, short-term upgrade, but it soon became clear that neither NEC nor the nuclear regulatory agency planned to shut down any reactors any time soon. Their decisions had the support of the government as it put state resources into modernization of Kozloduy rather than replacement energy. The plant supplied necessary power for domestic industry, which limped along in the wake of disastrous privatization programs, and covered an extremely inefficient residential consumption infrastructure that politicians were loath to change for fear of the ensuring voters' backlash. Elites faced incentives to forestall short-term costly policies in order to gain long-term benefits, yet Bulgaria did not collapse into either low rules stability or low compliance. Each level remained consistently in the middle due in part to the stability of the formal rules that conditioned the uncertainty in informal rules, as well as the strength of the regulatory agency to improve Kozloduy's safety and develop into a bone fide autonomous agency in the Bulgarian system.

The middle path

The rapid change of fortunes for BSP and SDS coincided with increasing economic hardships, rising social tensions over ethnic divisions within the country, and growing popular resistance to closing Kozloduy. Additionally, instability resulting from Berov's earlier resignation, and the ensuing confusion over which group would govern prior to the December 1994 elections, prevented the passage of necessary legislation to enable Western companies to begin safety work at Kozloduy, placing Bulgaria in jeopardy of violating the Grant Agreement toward the end of 1994.[18] These ills contributed to make SDS the scapegoat during the parliamentary elections and BSP easily gained nearly double the votes of the former government, prompting most of the SDS leadership to resign in disgrace. Yet despite the power the Socialists wielded given their parliamentary majority, the costs of full compliance remained as high for the new government as they had been previous ones, and the possibility of protest voting through early elections kept elites guessing as to how long BSP would govern. As a result, newly appointed NEC and Kozloduy personnel received instructions to continue their upgrades with the clear expectation that the work would ensure Kozloduy's long-term operation, not early closure a few years down the road.

The limitations facing the new Socialist government centered on public opinion of controversial topics like Western pressures on Kozloduy, but BSP had a strong mandate for solving the country's economic problems. With little opposition from the SDS President, whose only real power to veto the Socialist majority could easily override in Parliament, BSP Prime Minister Zhan Videnov set off on a course that nearly destroyed Bulgaria's economy and international reputation. The structures of government enabled BSP to push its agenda through Parliament but their failure to address mounting problems caused even greater hardships than those imposed by SDS's earlier privatization and austerity measures. During the next two years Videnov's government brought the country to its knees and permitted organized crime to penetrate vast regions of the government.[19]

Prime Minister Videnov also delayed implementation of the Grant Agreement by refusing to raise electricity prices to repay part of the loan. Prices had not increased by late 1995 with very little indication that they would rise in the near future. One safety expert said in September 1995 "the difficulty of implementing the initial NSA grant at Kozloduy does not augur well for the bigger safety improvement programs. It will have taken five years to install about 40 valves at Kozloduy. How long is it going to take to do the big backfit?"[20] The bulk of the safety work had not begun and its completion date began to stretch farther into the future. Meanwhile, the state utility and nuclear regulator agreed to restart unit 1 in late 1995 with an essentially unchanged set of reactors, despite serious concerns about the strength of their pressure valves. Regulatory chairman Yanev justified the decision and accused the West of violating the spirit of the Convention on Nuclear Safety, which states that all decisions regarding the operation of domestic nuclear facilities will be undertaken by the host country and are not subject to international controls. He would have been correct if the state of Bulgaria had not explicitly committed to early closure in the EBRD Grant Agreement, regardless of the party in power that did it.

Yanev's restart decision caused immediate problems. The French company EDF pulled out of the country after almost single-handedly maintaining a working relationship with the wayward Bulgarians. Yanev went ahead with the restart anyway, stating that the regulators would "shut it the moment we have information that it isn't safe. But I can't accept that any country pronounces a diktat and holds political press conferences. They claim that what the Russians have done [to test the valves] is not correct. They say 'it's not enough; we feel there should be more information.' But I cannot regulate on the basis of feeling."[21] When asked about the obvious connection between the valve issue and Bulgaria's commitment to shut down the reactors by 1998, Yanev said, "That's not my business."[22] He argued that the pressure valves would be safe enough to allow the plants to run until 2002 or 2003, regardless of any prior commitment. Even if that had not been the case, Yanev and his team were subject to the political vagaries of the time, which made long-term benefits of compliance seem less important than short-term costs.

The unstable informal rules of the game appeared even more clearly when the government fired Yanev "without warning" shortly after he issued the restart order.[23] The Socialist government seemed to split on this part of compliance as a part of the larger East-West discussion in the upcoming presidential election. The government's position had become very unstable due to massive social dislocation resulting from its half-hearted economic reforms, and threats of political violence began to appear as SDS and BSP tried to take control through new, extra-constitutional means. Making matters worse, President Zhelev called for an end to the parliamentary system in favor of a presidential one. Even though his calls went unheeded, the President's efforts added further uncertainty to an already confused political environment. While the Constitutional Court quickly and easily rebuffed Zhelev's efforts, instability in the informal rules of the game remained a serious problem.

Ladchezar Kostov took over as the new chief regulator in this context, bringing the requisite technical expertise to do the job, but like Yanev before him, Kostov took his cues from political elites struggling to gain their footing in the tumultuous political system, thereby ensuring the problems with Bulgarian compliance would continue despite the change in regulatory leadership. As an early example, Kostov authorized testing of unit 1's pressure vessel in late 1996 to show that they would last at least until the end of its operational life in 2004. Regulators did so as an attempt to de-legitimize the reasons behind the early closure commitment.

The political situation continued to affect compliance in general due to persistent instability in the informal rules of the game. As the Socialists presided over an ever-worsening economic crisis in 1997 and early elections loomed on the horizon, BSP leaders replaced Kostov with the more pro-environmental Gueorgui Kastchiev. The latter tended to fall more on the green side of the political spectrum by favoring a reduction in fossil fuel consumption while continuing to operate Kozloduy. The BSP leadership may have hoped his appointment would bolster popular support for their continued governance by setting him up as a Western-leaning candidate, ameliorating the growing resentments toward the socialist economic model. However, Kastchiev's appointment did not signal the most important change regarding the operation of Kozloduy. A major shakeup in the state utility also coincided with a renewed emphasis on long-term safety upgrades designed to keep Kozloduy running well into the future. The work being done with Western money had been allocated for short-term upgrades prior to closing the plants in 1998, but as the Bulgarians continued to delay initiating plans for replacement power and energy sector reform, the upgrades became a means of continued operation. Early shutdown had become far too dangerous politically, and no elite realistically wanted to pursue it since doing so would only diminish legitimacy in the eyes of a very Kozloduy-friendly populace. Citing the mood at the time, Kastchiev asserted in late 1997 that "the safety upgrading will never be finished."[24]

The seedy underside of Bulgarian politics

Corruption added to the uncertainty facing Bulgaria's elites, and increased instability in the informal rules of the game. While not to blame for its emergence, the BSP government of the mid-1990s allowed corruption to flourish in government ministries. Non-state actors exerted considerable influence as to how Bulgaria would be governed, and corruption made the delegation of power in the cabinet and state bureaucracies unclear. Both factors enabled repeated challenges to existing divisions of ministerial authority, and the harmful effects of corruption extended beyond the government directly as well. Because the constitution allowed for irregular election cycles, corruption became a key tool for punishing the party in power as a public whip to galvanize ire directed against opponents. The electorate had an easy time accepting those charges given the widespread abuses begun during the early SDS privatization schemes, whose effects continued throughout later governments.[25]

Three groups seized the opportunities presented by the first SDS government's attempts to change the state's role in the economy. State-sponsored athletes (the "wrestlers"), state security agents, and former *nomenklatura* effectively penetrated local businesses, customs bureaucracies, and privatization schemes,[26] while many political elites also maintained financial interests in privatizing companies.[27] Later anti-corruption campaigns curbed some links between organized crime and elected officials, but the problem remained a serious hindrance to effective governance.[28] Even though political elites followed the formal rules of the game, doing so usually amounted to "either a prelude to or an attempt of looting of state resources."[29] Corruption went beyond finding loop-holes in the law, and included licensing private banks to operate with state money, fraudulent privatization schemes, and creating a system for granting tax exemptions that benefited friends of the government. The legacy of those corrupt dealings often outlasted the politicians who committed them while general voter anger about corruption persisted across governments. Most importantly, whether those concerns had a basis in actual fact or merely perception by the public, the effect was still the same for elections as protest voting increased.

Corruption had been at its worst under the 1994 BSP government, which pursued a disastrous course of economic mismanagement and outright theft of state revenues. Living wages plummeted, inflation and unemployment soared, and the county seemed to drift internationally, neither fully committed to the Western sphere nor to an exclusive alignment with Russia. As a result of its failures and blatant corruption, mass anti-government demonstrations in January 1997 precipitated that year's pre-term elections, and SDS soundly defeated the Socialists and gained a considerable majority in Parliament. The impact on the rules of the game was more than simply changing the players though. Most disturbing to decision making was the lack of a predictable timeline to delineate election cycles between contests. The duration of the parliamentary election cycle ranged from one to three years, and even though presidential contests occurred more regularly, real power centered in Parliament, and presidential elections became more of an opportunity to vent rather than a source of stability in the informal rules of the game. Neither could elites count on repeat election victories between presidential and parliamentary elections. The massive swings of public support and the speed at which parties lost it made effective governance very difficult to achieve, further de-incentivizing long-term thinking. This unpredictability compounded the existing problems resulting from changes to bureaucratic responsibilities and widespread corruption. This situation did not favor any particular party as whoever was in office received all the blame for the country's troubles, and were summarily removed from office at unpredictable times only to have the process start over again the next time.

However, pre-term elections did not create instability by themselves; rather, unpredictability in how often they occurred created the uncertainty that so damaged Bulgarian politics.[30] Uncertainty shortened elites' time perspectives, shorter than they would have been given the stability in the formal structures of government. Consequently, shorter-term thinking prevented elites from pursuing the most costly aspects of compliance due to the inherent risk of pre-term elections. Quite understandably elites reasoned, why pursue something wildly unpopular that could only further limit one's chances of being in office for the full four-year term, to say nothing of being around longer? Thus while the formal rules functioned smoothly to protect the institutional power of each branch of government, the constitution also made it too easy to remove ruling politicians without giving them full opportunity to serve out their terms. The informal practice of back room bargaining and quid-pro-quo policy tradeoffs became extremely difficult under those conditions, and compromise itself became increasingly unlikely as the future offered little hope of recouping short-term concessions.

Exemplifying those problems, opposition members would walk out of Parliament to prevent the passage of government legislation. The practice was widespread and not limited to a few disgruntled MPs or even a single opposition party; Socialists and Democrats alike hindered regular governance in this way. Yet its occurrence throughout the 1990s followed no predictable pattern of when it would occur, thereby adding additional uncertainty to governance procedures. In an attempt to get around the problem, the 1994 BSP government changed the procedures for ministerial appointments and bureaucratic responsibilities to bypass parliamentary procedures. These could have become established practices but the rules changed again once the Democrats swept them from power in 1997.

With more than a two-to-one majority of seats, SDS had the chance to push through its legislative agenda nearly unopposed. This created another opportunity to change the way government worked by disrupting the previous BSP government's connections to organized crime, and initiating new government procedures for appointing ministers and defining their responsibilities. Some Socialist policies were left alone by the new government, primarily the social welfare programs, but by and large SDS discarded most of their predecessors' foreign and economic policies. While these actions moved the country closer to Western ideals, actual policy differences mattered less for nuclear compliance than did changes in policy-making procedures. The ways in which government functioned changed with each successive turn at the wheel, and uncertainty about the pattern of policy reversals continued.

SDS's successes showed in the general improvement of the economy, and the party moved the country closer toward EU integration and legal harmonization with the *acquis communitaire*, the nearly eighty thousand pages of regulations required for membership. BSP challenges to the government's policies and procedures continued unabated despite their obvious successes. Several attempted votes of no-confidence did very little to harm the government's position, but they did impact the rules of the game by helping to maintain uncertainty regarding the "basic legislative acts regulating governing procedures."[31] The damage was limited to the informal rules as "disputes over jurisdiction, expansionist institutional strategies, and truculent usurpation of authority [were] rare occurrences."[32]

Attempts to subvert the constitution occurred at times but were generally limited in number and the scale of the effort. The Socialists tried twice to curtail the President's appointment powers by making them contingent on parliamentary approval, and President Zhelev had earlier proposed a stronger presidential form of government.[33] Yet in all cases the formal rules survived unchanged. Even though respecting the rules did not translate to cooperation in policy making, "the rules of the game have so far kept the vituperative skirmishes between politicians from degenerating into a 'bare-knuckles' brawl between institutions."[34] The judiciary earned some of the credit for those limitations by effectively constraining parliamentary groups from usurping presidential powers, and helping to quash President Zhelev's attempt. Yet as the primary mediator in cases of those conflicts, the judiciary itself occasionally became the center of controversy as well, with challenges by both BSP and SDS focused more on the procedures defining its influence rather than the constitutionally defined independence of the courts.[35] This kind of formal stability in the judiciary did not translate to informal stability though,[36] as elites still contended with uncertainty about the durability of long-term policy agendas, as well as how long elites can reasonably predict they would be in office to ensure their completion.

Middling improvement

Despite the negatives facing the political system, a period of increasing stability in the informal rules of the game occurred in the late 1990s and compliance made some marked improvements during that time. The 1997 SDS government succeeded in curbing inflation, bringing it down from its zenith of over one thousand percent to a modest twenty-two percent within a single year, while real GDP showed positive growth and foreign investment jumped six hundred percent from the preceding year. Anti-corruption efforts also had a good beginning, although the problem remained considerable.[37] How they succeeded is less important for compliance than the fact that their successes gave good indications for an election victory in 2001. The future looked brighter for the ruling SDS government during its first two years as Western support increased and Bulgaria seemed both stable and cooperative.[38] More importantly, popular opinion aligned with SDS's pro-Western goals, something unheard of during the preceding six years.[39] All of this began to increase governance stability, and the declining threat of unpredictable pre-term elections meant that SDS could realistically expect their procedures to last for some time. Ministerial appointments became more durable and policy discussions less erratic giving the government's newfound freedom to pursue longer-term goals.[40] This enabled a longer view of costly compliance decisions in order to gain long-term benefits as a result.

SDS certainly moved in the right direct but the case should not be overstated. Successes in the economy and control of both the presidency and Parliament proved insufficient to prevent the return of widespread uncertainty in the political system. While outright failure can be the cause of electoral defeat in any democracy, in Bulgaria the necessity of always keeping a protest-laden electorate on one's side prevented sustained long-term policy making for the long-term. At best, during the first two years of SDS's term, certainty and stability in the informal rules of the game increased. However, this did not represent consistent progress toward increasing stability overall.

Thus while the SDS government approved a new energy bill in 1997 designed to bring the country's legislation more in line with EU accession requirements, it did not keep pace with requirements to improve the regulator's financial sufficiency, and even worse, remained adamant about not closing Kozloduy before the end of its design lifetime. Closure dates for units 1–4 were set for sometime between 2004 and 2012, noticeably later than the original 1998 dates established in Bulgaria's Grant Agreement, and almost identical to the units' scheduled end-of-lifetime. However, despite the obvious problems for compliance, Western officials viewed even this small step as a move in the right direction since it represented at least some recognition that Kozloduy had to close down sometime in the near future.[41]

Even with general reluctance for early shutdown among Bulgaria's political and nuclear industry elite, differences appeared between those favoring on time closure between 2004 and 2012 and those pursuing lifetime extensions beyond those dates. Either choice put Bulgaria in violation of the Grant Agreement, something confirmed in April 1998 as NEC Vice President for Research and Development Ivan Hinovski said Kozloduy 1–4 would undergo a "very extensive modernization with the goal of bring the units to international standards and operating them to the end of their design lives."[42] While Regulatory Chairman Kastchiev asserted NEC would have to show the plan met all safety requirements deemed necessary by his organization, little doubt existed that the regulator would stop the program. According to NEC and Kastchiev's arguments, safety had become a justification for rejecting Bulgaria's early closure commitments to the EBRD and the European Union, arguing "as long as the plants are safe they can and will be allowed to operate."[43] Western officials responded that they were prepared to hold up the Bulgarians' accession to the European Union if they continued to forgo their commitments.

Complicating matters, the Bulgarian government could not pay for safety upgrades on its own given the high costs and limited state resources. Western banks and governments staunchly refused to fund a plan that clearly violated the agreements they wanted to ensure the Bulgarians did not break in the first place. Western officials had previously used the long-term benefit of compliance – greater integration with the West through EU membership – as an incentive to increase Bulgarian compliance, but in each case the Bulgarians' short-term needs limited long-term thinking. They subsequently complied with some aspects but not the most costly ones.

As SDS dithered on decommissioning, invoking the increasing ire of the EU accession negotiators, public opinion began to turn. BSP had also regrouped and once again called for early elections in early 1999 since economic growth rates failed to match the Democrats' campaign promises. Holding steady throughout the tumultuous economic rollercoaster, electricity exports offered one of the few reliable sources of revenue for the state, and as a result, an increasing number of parliamentary voices began to call for lifetime extensions for Kozloduy, the primary energy producer in the country. To save their positions and avoid an early end to their time in power, SDS MPs began supporting opposition calls to run the plants as long as possible. Operational safety had improved dramatically during the 1990s so the decision at least had some technical backing. However, the upgrades did not cover long-term problems, and like Chernobyl, they kept the plant running at the minimum safe level while still allowing for the worst case to happen.

Improved safety became the justification for attempting to break the 1998 commitment as the plant became more efficient, enabling Kozloduy to export nearly fifty percent of its generated power to Southern Europe. Increased profitability sent much needed revenues to crucial social and economic programs, relieving the pressure on otherwise stretched national budgets. Medium compliance itself became profitable as the government dedicated more money to making the regulator less dependent on payments from the national utility or Kozloduy itself, and the regulator worked to keep Kozloduy profitable.

Trying to give a sense of legitimacy to their renegade behavior, Parliament approved a plan to renegotiate the EBRD Grant Agreement in March 1999, and unanimously rejected any shutdown before the end of the reactors' design lives.[44] Pro-Western Democrats controlled the Parliament so pro-Western policy alignment cannot explain this blatant rejection of the early closure commitment. While the game had achieved enough stability that elites could count on the system itself persisting regardless of who headed the government, persistent instability in the informal rules and the resulting uncertainty about future play in the political game framed elite considerations of the costliest aspects of compliance.

Not surprisingly, key Bulgarian officials rejected Western criticism for their failures to invest in alternative energy, reform the energy sector,[45] and shut down Kozloduy's first-generation VVER reactors before the end of the decade. Instead they maintained they were in full compliance because the terms of the agreement had been unacceptable from the start. Milo Kovachev, head of NEC's safety and operations divisions, claimed that there was "no formal statement from anywhere in the EU that closing reactors is an absolute precondition to start negotiations. Our theory is that we are sticking to our commitments."[46] The EC did not agree and conditioned membership negotiations on renewed progress toward high compliance.[47]

As discussions took on increasingly confrontational tones, former regulator chief Yanko Yanev expressed a common view held at the time. He repeatedly referred to the West's attempts to bankrupt Bulgaria by forcing the country to shut down its perfectly safe nuclear power reactors. He told me early closure demands were politically motivated and had nothing to do with safety. In his view "safety is determined by a given society, not some politician in another country that could never fully understand the needs of another country."[48] While he may have been technically correct about some aspects of Kozloduy's status, a ruling Bulgarian government had committed to early closure regardless of differences in safety assessments. It signed a binding agreement that contained no provisions for removing closure requirements based on improved safety.

Despite his rejection of the political nature of compliance, Yanev did not blame the West for the continued problems facing Bulgaria's nuclear regulatory authority, something many Western officials acknowledged.[49] However, Yanev complained that assistance money all too often went needlessly to completing reports that studied the need for more reports.

> I would say, "Why are you giving me all these papers?" Even in the beginning in 1991, when I was in Brussels to sign an agreement for crash assistance, $30 million, I looked and all the money went to study this or that. Where is all the money for real needed work? They said, "No, this is to study the need." I didn't sign it and went home, and the government said, "What the hell are you doing? You've created a big mess." I said, "I'm not going to sign it because there isn't anything for us." That was the problem.[50]

Yanev did not stand alone in his view that something other than safety lay behind the West's demands to shut Kozloduy,[51] and successive Bulgarian governments consistently attempted to get out of their collective arrangement with the West. Yet rather than opt out by refusing future assistance money, each government revealed the Janus-face of Bulgarian compliance by accepting the immediate short-term assistance while rejecting the costs necessary to ensure long-term cooperation with the West. The West had no choice but to continue the relationship for fear of Kozloduy operating without its ill-gotten safety upgrades; in the end, some progress was still better than none at all.

One bright spot in the overall negative pattern occurred in November 1999 when SDS Prime Minister Kostov announced that Bulgaria would shut down the first two units before 2003, and the last two sometime around 2006. As SDS maintained a strong position and its future prospects indicated a likely full term in office, Kostov signed a complementary Memorandum of Understanding with the EC to reinforce the commitment to comply with the spirit of the EBRD agreement, even if not its letter. Noting the past problems and the somewhat shortened lifetime plan for Kozloduy, the European Union compromised and invited Bulgaria to begin accession discussions if the Bulgarians kept to the MoU's terms.

Rival elites in Parliament and the nuclear industry began questioning those dates almost immediately though. The NEC board decided to forgo upgrades at units 1 and 2 as a result of the 1999 MoU early closure commitment as it saw no need for costly upgrades when the reactors would be closed in only a few years time. However, regulatory chairman Kastchiev countered that he did not think the reactors would close then and continued to push for their continued operation.[52] Prime Minister Kostov initially continued to push for compliance with the 1999 commitment despite intense opposition, but as time went by and the government's political fortunes began to grow dim over the next few months, more elites acknowledged that sticking to the 2006 shut down of units 3 and 4 was unlikely. Kozloduy's executive director confirmed in June 2000 that 1 and 2 would close on time, but he said "plant management is going ahead with modernizing 3 and 4 in a bid to operate them through their design lifetimes."[53]

Even the Prime Minister himself eventually expressed reservations about his 1999 decision as concerns grew about pre-term elections and protest voting. A year after his November announcement, Kostov claimed that the MoU conditioned the shutdown of 3 and 4 on receiving sufficient financial assistance from the West to pay for the decommissioning of the first two units. He wanted to run the newer units until 2008–2010 respectively, still short of their design lives but also beyond the agreement dates set with the EC. A year had gone by and once again Bulgaria's elites continued to deny the need to comply fully with their international agreements due in large part to the persistent uncertainty in the actual governance of day-to-day politics and the resulting short time horizons. Formal stability gave a measure of certainty since elites did not have to worry about changes in the constitutional framework, but short-term concerns carried a lot of weight for elites as they tried to establish some form of authoritative government under conditions of informal uncertainty.

The unraveling of SDS's strategy compounded those concerns in 1999 as two years of renewed privatization continued to be a source of corruption in the government, and austerity measures bit deeply into the country's economic progress. Growth slowed and inflation rose sharply in 2000. Despite considerable efforts to stabilize the economy with "comprehensive macroeconomic and structural reforms, the introduction of a currency board, an intensified privatization process, and a revamped tax system, these changes had an increasingly negative impact on family budgets."[54] In the end SDS became the scapegoat as once again voters punished the party in power.[55] With the scheduled elections at hand, BSP opted to wait and did not call for a vote of no-confidence; they would likely have called for early elections if the timeline had been longer. As the chances for a repeat election victory dwindled, SDS elites once again began to adopt shorter-term goals in an attempt to win back popular support. This had an immediate effect on compliance with nuclear safety commitments as elites across the political spectrum retrenched and protected Kozloduy as a treasured national asset. Yet even the attempted rapprochement with the populace did not have time to take root as a political wild card entered the scene and changed the nature of the political process once again.

Janus replaced?

In April 2001, three months before the first full-term parliamentary election, voters faced a new choice as former King Simeon II entered the political scene with the creation of a loosely organized, vaguely articulated policy agenda fittingly called the National Movement for Simeon the Second (NDSV).[56] Like a breath of fresh air Simeon's party swept through the country with claims of increasing women's participation,[57] ending corruption, and revitalizing the economy in eight hundred days.[58] NDSV won a landslide victory over SDS, gaining fifty percent of the seats in the National Assembly. After falling one short of an outright majority Simeon entered into a coalition with the minority DPS Party responsible for breaking up SDS's first governing coalition nearly ten years before.

While several BSP members accepted seats on Simeon's team, SDS remained in staunch opposition and refused any part in the government even though NDSV supported the incumbent SDS candidate, Petr Stoyanov, in the November 2001 presidential election. Unfortunately, broad, high hopes placed on the novelty of the monarch's presence quickly wore off, and six months after Simeon's victory, public support degraded so much that even the popular Stoyanov could not survive against the BSP candidate as the Socialists presented themselves as the only alternative to the SDS/NDSV presidential candidate. Thus, even though Stoyanov had several things going for him – he was the most trusted national politician, he had the backing of the Prime Minister, and he held a sizeable lead over Georgy Parvanov in opinion polls leading up to the election – Stoyanov still narrowly lost in the second round of voting. Once again no Bulgarian party could hold on to power in two consecutive elections, and procedural instability continued from government to government.

In that context, Simeon fired Regulatory Chairman Kastchiev shortly after the presidential election, and replaced him with the former CUAEPP deputy director Robert Popits. The new NDSV government initiated legislation to give the nuclear regulatory authority full autonomy and financial resources to do its job, a process that did not end until mid-2002. Some progress also continued on decommissioning units 1 and 2 as the EBRD agreed to nearly $265 million in assistance conditioned on Bulgaria's compliance with the 2006 early closure commitments for units 3 and 4.

The year 2006 became a critical date because the European Union made it clear that no first-generation Soviet reactors would enter the Union. Since Bulgaria initially hoped to get in by 2007, the date for decommissioning had been set at the latest possible moment, but still earlier than the Bulgarians would have liked. However, a serious contention arose among Bulgaria's elites when Lithuania and Slovakia both received invitations to begin accession talks that granted admission dates several years prior to their closure dates, a decision Yanev labeled as completely unfair.[59] EU officials responded to increasing Bulgarian complaints by countering that the other two countries had made much more progress on the related issues necessary to ensure closure, issues that the Bulgarians had consistently balked at over the years – raising prices and setting aside a decommissioning fund.

Simeon's government maintained the previous SDS policies regarding Kozloduy because it too operated in an environment of unstable informal rules within the context of stable formal rules. It remained difficult to establish clear, predictable procedures for governing the country despite the Prime Minister's early claims to the contrary. This problem continued to create unwillingness among Bulgaria's ruling elites to guarantee long-term commitments. Indicating this view did not exist solely in the old parties in Bulgaria, Simeon himself said after a visit to the plant that he thought units 3 and 4 should operate longer than 2006, and he would try to convince the European Union to change its position. Indicating that rival groups in Parliament and the government continued to vie for control over compliance decisions, former Energy Minister Roumen Ovcharov blasted Simeon by saying he did not have that kind of power.

For a brief period under SDS that competition had appeared to diminish, only to resurface a year later despite a new party in power. The problem remained unresolved as evidenced when Energy Minister Milo Kovachev stated publicly that Kozloduy 3 and 4 would operate beyond 2006 to ensure the energy balance of the country and the region. He and former Energy Minister Ovcharov wanted to run them until 2010–2012.[60] The new nuclear regulator, Emil Varipev went a step further saying that there was "no justification for shutting Kozloduy 3 and 4 in 2006, or even 2010–2012," accusing the EC of using outdated information "from the early 1990s when the plant was in bad shape. But the situation has changed dramatically since then. The so-called design lifetime assigned to Soviet reactors is an outdated idea. I can't say what the predicted life of the units would be. It could be 2020 or even 2024."[61]

Then in mid-2002 Bulgaria's political and nuclear officials began their move to keep units 3 and 4 running beyond 2006. After an IAEA Safety Review team concluded in June that the reactors had reached a comparable safety level to Western plants, additional supporters from Slovakia, Euratom, and the World Council of Nuclear Workers joined the effort to keep Kozloduy online.[62] In addition to the Chairman of the nuclear regulatory authority, Kozloduy's new manager Iordan Kostadinov said he wanted to pry Bulgaria loose from its commitment to the European Union to shut the units 3 and 4 by 2006.[63] He supported the Bulgarian energy law at the time that envisioned shutdown for 2008–2010.

BSP President Purvanov brought out an even bigger weapon for the fight when he sent a letter to the leaders of all the EU member states. He claimed that if the European Union forced Bulgaria to comply with its shutdown commitments social unrest would become so great that the country's move toward greater democratization would be jeopardized. He referenced nearly seventy percent of the population thinking the "Bulgarian government should resolutely insist on postponing the decommissioning of the 1st and 2nd Kozloduy units even at the cost of delaying entry in the EU."[64] The issue was a political time bomb waiting to explode on whichever government was in power at the time.

Purvanov skillfully played to the crowd and had been an ardent opponent to shutting any reactors at Kozloduy. In July 2002 he asserted that "Kozloduy 3–4 will not be closed by 2006 as the EC wants, and that units 1–2 will be kept closed for the time it takes to modernize them,"[65] calling into question the terms of compliance once again. This kind of blatant unwillingness to comply existed across the three major parties. Veselin Bliznakov, NDSV Chairman of Parliament's energy committee, and Vladimir Kossiev, deputy chairman of SDS, also voiced their disapproval with early closure. Kossiev called it the "worst possible mistake" given what Bliznakov called the "religion" of nuclear power in Bulgaria.[66] Their statements indicated more than a problem for compliance directly; they also showed contending views about which government authority had the right to make the final decision on Kozloduy's fate.

However, as with the other areas of Bulgarian compliance, while various government and nuclear officials criticized and threatened to make plans to keep the plant running, decommissioning plans proceeded to a limited degree as Kozloduy plant management received bids for the Project Management Unit in early September 2002, the group responsible for running the shutdown work. The Prime Minister also said Bulgaria would abide by the 1999 agreement concerning units 1 and 2 and close them by year's end. Then later that year, in September the Bulgarian Council of Ministers officially accepted the 2006 date for closing 3 and 4, but asked for a safety review to prove that the date was inappropriate since they considered the plants safe enough to continue running past the deadline. Bulgarian European Policy Minister Meglena Kuneva added that "if the [European] Commission does not agree to a Kozloduy safety review, we won't reach agreement and sign the energy chapter."[67]

SDS and the left-wing Coalition for Bulgaria (dominated by BSP) responded by filing court motions that Simeon's government had overstepped its bounds and threatened a vote of no-confidence. He held a fragile majority in Parliament and almost eighty percent of the population did not want any part of Kozloduy closed at all,[68] so in the face of such overwhelming popular pressure and the weakness of its hold on power, NDSV's governing coalition voted in early October 2002 to reject closing units 3 and 4 before Bulgaria joined the European Union.

Not only did the procedures governing the country remain uncertain due to the presence of protest voting and threats to shorten election terms, uncertainty concerning Kozloduy remained high as different groups claimed authority over its fate. Under such conditions of uncertain governance within a stable formal structure of government, it is reasonable to assert that elite time perspectives remained short enough to reject the most costly aspects of compliance, while long enough to foster agreement with those measures that could be more easily justified and endured by the electorate.

Conclusion

Given the longstanding duplicity in Bulgarian compliance, why then did Kozloduy's first two units close in 2002 in line with the later MoU guidelines, and the more contentious 3 and 4 follow in 2006? What enabled the final decision to cease operations before the end of the reactors' design lives? Did the Bulgarians reach high compliance, or like the case in Ukraine, did they simply have no choice as the West called their bluff?

It is challenging to explain the different degrees of Bulgarian compliance. Elites committed, reneged, recommitted, reneged, and finally shutdown Kozloduy's first four reactors before entering the European Union. They did so after complying with the CNS and Grant Agreement conditions for medium compliance by creating and funding the new Nuclear Regulatory Authority (NRA);[69] Kozloduy had already undergone extensive safety upgrades to meet the conditions for low compliance. Yet even NRA joined the cacophony of voices claiming that units 3 and

4 did not fall under the original classification of first-generation Soviet-designed 440-230s, and therefore should not close in 2006. Yanev told me that they wanted them reclassified as a hybrid model of the VVER 440 since the plant had undergone tests for pipe rupture scenarios that showed it could withstand a break in a five hundred millimeter pipe and still prevent damage to the reactor core. This would have meant that units 3 and 4 did not fall under the G7 decision from his perspective, and therefore should not have been not subject to the terms of Bulgaria's existing international commitments. From Yanev's comments and the efforts made by the regulatory agency, clearly different concepts of safety were at odds with compliance, but while the Bulgarians hotly contested the safety justifications for the original EBRD Grant Agreement and the MoU with the EC that followed six years later, the fact remained that successive Bulgarian governments signed those agreements and ultimately closed Kozloduy's four oldest reactors before the end of their design lives.

Unpredictability in compliance behavior resulted from the lack of comprehensive stability in the political rules of the game. Medium rules stability defined Bulgaria's compliance period leading elites to make contradictory and often short-lived sacrifices. The stability of the formal rules enabled elites to function within a known structural framework; the constitutional provisions have been clear, durable, and predictable. Yet beyond that assurance, elites framed their evaluations based on the realities of politics in Bulgaria: the chances of being around for the long-term remained slim as the overwhelming majority of governments ended before their full terms. Uncertainty about future opportunities to implement legislation, recover losses, and gain long-term benefits made costly short-term decisions like decommissioning Kozloduy very unlikely for most of the time. Elites had incentives to do as little as possible to precipitate the inevitable votes of no-confidence and subsequent early elections. Consequently, it is reasonable to assume that elite time horizons were shorter than they would have been if the country operated under both stable formal and informal rules, thus making some compliance decisions acceptable, but preventing full compliance when the electorate refused it.

In contrast to this explanation, Western officials I interviewed often deplored the lack of safety culture in their Bulgarian counterparts, using that difference to explain levels of compliance. They claimed that if only the political and nuclear officials understood the inherent risks of operating a nuclear reactor without a containment structure they would have readily accepted the need to comply more fully. Yet the Bulgarians clearly did not want to shut down anything at Kozloduy, believing all the reactors to be safe, something later IAEA reports argued as well. While politics between the European Union, the United States, and the IAEA shed some light on why the first two disregard the assessments of the third – throughout most of its history the West claimed the IAEA operated as a pawn of Russian interests – the real problem is compliance, not the unique merits of competing safety assessments. A safety culture argument cannot explain why elites with a risk acceptant culture would make costly safety improvements when that money could have gone to help resolve the country's pressing economic woes.

In addition, the Bulgarians increased the autonomy and authority of the nuclear regulator yet safety culture remained consistent as nuclear officials held that their view of safety was correct. Such a view stood in opposition to the one espoused in Bulgaria's international agreements, and the model presented here states that the Bulgarians closed the four units at Kozloduy in order to get into the European Union, not because of a change in safety culture. However, that fact in itself does not explain why they actually did so given the risks the Bulgarians took by consistently arguing against the requirement. The European Union made clear the possibility of passing over Bulgaria if Kozloduy's four older reactors remained operational. The incentive mattered, but the hard-line position had not wavered; the Bulgarians faced firm commitments and clear requirements throughout the compliance period.

A counter-argument posits that the shadow of 1998 may have extended into their collective thinking so that Bulgarian elites envisioned a similar retrenchment by the West in the mid-2000s. However, this explanation of changing international incentives misses the fact that the MoU with the EC referenced 1998 negatively, and for several years the West held an immovable position requiring early closure. Therefore the explanation for the 2006 decision to shut down 3 and 4 lies elsewhere within the elite decision-making process conditioned by rule stability.

As with SDS's earlier period of increasing certainty that enabled elites to foresee a realistic chance for longer play in the game, so too NDSV had a greater degree of stability in which to operate as its term in office drew closer to completion. Riding a strong election mandate, the government had closed units 1 and 2 without the anticipated public uproar and dire economic collapse as the European Union courted the country with the incentives of membership, and the populace grew more receptive and positive about the benefits.[70] Since NDSV could realistically count on being able to fulfill its term in office, the government had enough opportunity to see accurately the requirements facing the country – comply and join the European Union or repeat the past and get passed over. As a result, safety arguments faded to the background as NDSV gambled on a brighter future in the European Union supported by an increasingly Western-leaning electorate, despite its general reluctance to sacrifice Kozloduy.

However, NDSV lost the ensuing parliamentary election in 2005 to a center-left coalition despite polling closely to BSP beforehand. The Socialists should have followed their normal course and rejected the compliance requirement to close 3 and 4 in 2006, except for one additional factor that affected elite time horizons: the lowest voter turnout since Bulgaria emerged from one-party rule. After ten years of protest voting, it appeared that the Bulgarian people were beginning to tire of the political rollercoaster and had settled on pro-Western economic goals, which they could more easily maximize in the European Union as evidenced by the successes of Poland, Hungary, and the Czech Republic the year before. Thus, with less potential pressure of protest voting and lower threats to their term in office, even BSP could see that the long-term benefits of compliance outweighed the short-term costs. In the end then, a decline in democratic efficacy helped the country close the accession protocol as BSP agreed to close Kozloduy's last two unsafe reactors.

The model presented here shows that rules stability conditioned Bulgaria's compliance behavior. It accounts for the inputs into that model, but the process itself matters most. As a result, unlike Ukraine, which would have continued operating Chernobyl if not for its lack of fuel and financing, Bulgarian elites made the decision to shut down their nuclear reactors before the end of their design lives. However, they did so after fundamentally violating the terms of the Grant Agreement with the EBRD. In order to maintain a consistent metric, Bulgaria must receive a score of medium compliance due to the original failure to decommission Kozloduy 1–4 in 1998. However, while this matters in terms of the model's internal consistency, the fact remains that Kozloduy 1–4 closed, and the Bulgarians deserve credit for closing four of the most dangerous reactors in the world, even if it took a long time and followed an unpredictable route getting there.

Table 5.1 Bulgaria Independence (1990)–Decommissioning (1 and 2: 2002, 3 and 4: 2006)

Rules Stability	*Clarity*	*Predictability*
Formal	Yes	Yes
Informal	No	No
Conclusion: Formal rules experienced no serious threats to the constitutional delegation of power or transfer mechanisms, despite occasional partial threats to change the system. Informal rules remained unpredictable for most of the period, with two significant periods of increasing rules stability due to predicted full terms in office and later, declining threats of protest voting due to lower voter involvement in the election cycle. Bulgaria indicates **MEDIUM STABILITY**.		
Compliance	• Completed safety upgrades • Sufficient legal autonomy and funding for the NRA, increasing personnel, and operation within parameters of CNS • Delayed compliance with Grant Agreement, revised timeline, compliance with later MoU – units 1 and 2 closed on time in 2002, units 3 and 4 in 2006	
Conclusion: Bulgaria indicates **MEDIUM COMPLIANCE due to changed conditions of compliance**.		

Notes

1 Bell (1997): 353.
2 See Kolarova (1999): 152 for general trends in the first decade.
3 See Ganev (2001): 211 for greater detail about the causes of governance instability.
4 Ibid., 191.
5 Ibid.
6 *Nucleonics Week*, February 4, 1993: 1.
7 Fire danger was just the "tip of the iceberg" according to Thierry Baudon, director for Public Infrastructure, Energy and the Environment at the EBRD. In an interview with Nucleonics Week, he said the "the real problem is that it is poorly managed, badly run, and could run into trouble every day. Today it's the generator, tomorrow it could be a primary circuit leak" (*Nucleonics Week*, January 1, 1993: 1).

8 Carol Kessler, in discussion with the author, September 2002.
9 This put Bulgaria in non-compliance with the Convention on Nuclear Safety beginning in 1994. That would not change until 2001 when the regulatory agency finally became autonomous, although limited steps toward that end occurred between 1997 and 1999.
10 Yanko Yanev, in discussion with the author, September 2002.
11 *Nucleonics Week*, May 13, 1993: 1.
12 Yanko Yanev, in discussion with the author, September 2002. According to one Western expert who was involved in the Kozloduy enterprise, "come what may, if Kozloduy's power is needed, the units will continue to operate. If we help them, they'll improve the plant, but they'll run it in any case. When you come right down to it, the Bulgarians are saying, 'if you want to help us, great. If you don't, too bad.'" *Nucleonics Week*, June 24, 1993: 1.
13 Ibid.
14 Ibid.
15 *Nucleonics Week*, June 24, 1993: 13.
16 *Nucleonics Week*, December 9, 1993: 10.
17 Ibid.
18 The Agreement stipulated that all equipment had to be installed and upgrades progressing by 1995. Very little work had been done by late 1994 making it unlikely the schedule would be kept.
19 Taras (1998): 115.
20 *Nucleonics Week*, September 21, 1995: 1.
21 *Nucleonics Week*, October 12, 1995: 1.
22 Ibid.
23 *Nucleonics Week*, September 12, 1996: 5.
24 *Nucleonics Week*, December 11, 1997: 1.
25 A good example was the 1995 incident when Prime Minister Videnov let several of his close friends export large quantities of wheat despite government restrictions. His actions led to severe domestic shortages for several weeks; see *East European Constitutional Review* (1997).
26 Ibid.
27 Ibid.
28 The European Commission's *Regular Report on Bulgaria's Progress towards Accession* (1998): 9, states that "Corruption remains a serious problem in Bulgaria, [and that] despite measures and a substantial increase in public awareness, considerable further efforts are needed in order to achieve tangible results."
29 Ganev (2001): 209.
30 While the Bulgarian election mechanisms have some similarities to those in the United Kingdom and Germany, Bulgarian pre-term elections have not been helpful given their negative impact on the establishment of authoritative governments capable of ensuring stable informal rules of the game necessary for full compliance with nuclear safety agreements.
31 Kolarova (1999): 155. SDS used the same tactic against the later NDSV government. This points to unstable informal rules, not the particular policy strategies of a single political party. Opposition parties by nature oppose the ruling government, so neither BSP nor SDS added to the problem, they just perpetuated it. See *East European Constitutional Review* (1998) for more specific details.
32 Bell (1997): 194.
33 Ibid., 191.
34 Ganev (2001): 195.
35 Assessments are given in Karatnycky, Motyl, and Graybow (1999): 163; Karatnycky, Motyl, and Piano (2000): 183; Karatnycky, Motyl, and Schnetzer (2002): 140; (2003): 133; (2003) and European Commission (1998–2002).
36 Ganev (2001): 210.

37 *East European Constitutional Review* (1997) gives an overview of the initial steps taken by SDS in 1997. European Commission (2001) details the marginal successes resulting from those efforts.
38 European Commission (1998–1999) for evidence of the change.
39 See Kolarova (1999) for more details on public support.
40 SDS won the 1996 presidential election giving them unopposed power for the first time. The Socialists had had similar control for two years from 1994 to 1996 when Petar Stoyanov (SDS) was elected President. I do not mean to imply this enabled either side to escape the instability in the informal rules resulting from early elections and hostile opposition attempts to derail government policies. It is simply to say that at that time things looked good for SDS to win the next election.
41 Carol Kessler and David Sycamore, Director for Institutional Affairs, Foratom, in discussion with the author, September 2002. Ms. Kessler indicated it was progress toward closing Kozloduy and Mr. Sycamore countered that represented a more accurate understanding of the plant's true safety status.
42 *Nucleonics Week*, April 23, 1998: 7.
43 Ibid.
44 *Nucleonics Week*, March 18, 1999: 13.
45 Energy demand had fallen far short of earlier predictions, and fundamental infrastructure changes had not been accomplished. Most buildings in Sofia used centralized heating that did not allow individual customers to control the temperature in their apartments. As a result, many would simply open windows to cool down their rooms in the winter. In addition, demand in general remained low due to weak commercial activities, and the claims that Kozloduy was necessary to meet domestic needs carried little weight with Western officials.
46 *Nucleonics Week*, September 16, 1999: 12.
47 *Nucleonics Week*, October 21, 1999: 18.
48 Yanko Yanev, in discussion with the author, September 2002.
49 See Western European Nuclear Regulators Association (2000) and the Council of the European Union (2001) for important examples.
50 Yanko Yanev, in discussion with the author, September 2002.
51 The IAEA and Foratom have regularly backed up the Bulgarian claim that Kozloduy was safe and did not need to be shut down. That may or may not be true. The central issue for my project is compliance, not the validity of safety standards.
52 The legal autonomy of the regulatory body was still in doubt as Kastchiev's decision should have been the final word if Bulgaria was operating under the terms of the Convention on Nuclear Safety (1994).
53 *Nucleonics Week*, June 22, 2000: 6.
54 Karatnycky, Motyl, and Schnetzer (2003): 127.
55 For a good study on voter behavior and the weakness of Bulgarian civil society, see Nagle and Mahr (1999).
56 The Constitutional Court ruled that he could not run for President as he originally intended since the constitution requires all candidates to live in Bulgaria for a minimum of two and a half consecutive years out of the preceding five. Simeon settled for going after the prime minister's office instead.
57 He had been forced earlier to create a coalition with the Bulgarian Women's Party and the National Movement for Revival "after a Sofia City court refused to register his movement since it failed to meet certain requirements" (Inter-parliamentary Union, 2002–2012).
58 This was an interesting move to give himself two years to work out the kinks in his agenda – enough time to quell short-term criticism, but still not so long as to minimize the immediacy of his plans when he presented them to the public.
59 Yanko Yanev, in discussion with the author, September 2002.
60 *Nucleonics Week*, January 31, 2002: 1.
61 Ibid.

62 Broader divisions between EC and US safety officials aside, the IAEA's assessment did not directly affect the issue of compliance since the West claimed it as binding on Bulgaria. More damaging to that view, World Council of Nuclear Workers (WONUC) brought suit against the European Commission claiming that the MoU agreement in 1999 was in fact non-binding, and the EU court ruled that no legal grounds existed for forcing closure in any EU candidate country. The legal opportunity not to comply did not translate to forcing the European Union to let Bulgaria in though, and as a result, compliance remained the central issue (*Nucleonics Week*, August 15, 2002: 1).
63 *Nucleonics Week*, July 4, 2002: 1.
64 BBSS Gallup International (2002).
65 *Nucleonics Week*, 8/15/02: p1.
66 *Nucleonics Week*, July 11, 2002: 1.
67 *Nucleonics Week*, October 3, 2002: 1.
68 Ibid.
69 Bulgaria Nuclear Regulatory Agency (2002–2012).
70 European Union (2006).

References

BBSS Gallup International. *Perceptions of the Issue of Decommissioning Kozloduy N-Plant Units, March–May 2002*. 2002.

Bell, John. "Democratization and Political Participation in 'Postcommunist' Bulgaria." In *Politics, Power, and the Struggle for Democracy in South-east Europe*, edited by Karen Dawisha and Bruce Parrott, chapter 9. New York: Cambridge University Press, 1997.

Bulgaria Nuclear Regulatory Agency. *Rules of Procedures of the Nuclear Regulatory Agency*. Accessed regularly 2002–2012. www.bnsa.bas.bg/en/documents-en/legislation/rulesofproced/rules-of-procedure.pdf.

International Atomic Energy Agency (IAEA). Convention on Nuclear Safety. 1994.

East European Constitutional Review 6, no. 4 (Fall 1997); 7, no. 1 (Winter 1998).

European Commission. *Regular Report on Bulgaria's Progress towards Accession*. November 1998.

——. *Regular Report on Bulgaria's Progress towards Accession*. October 1999.

——. *Regular Report on Bulgaria's Progress towards Accession*. November 2000.

——. *Regular Report on Bulgaria's Progress towards Accession*. November 2001.

——. *Regular Report on Bulgaria's Progress towards Accession*. October 2002.

European Union. *Eurobarameter 65: Public Opinion in the European Union*, Autumn 2006. Accessed January 2007. http://ec.europa.eu/public_opinion/archives/eb/eb65/eb65_bg_exec.pdf.

——. *Report on Nuclear Safety in the Context of Enlargement*. 9181/01 LIMITE ATO 36 ELARG 118. Brussels, Belgium. May 27, 2001.

Ganev, Venelin. "Bulgaria: The (Ir)Relevance of Postcommunist Constitutionalism." In *Democratic Consolidation in Eastern Europe: Institutional Engineering, Volume 1*, edited by Jan Zielonka, chapter 7. New York: Oxford University Press, 2001.

Inter-parliamentary Union. "Bulgaria National Assembly." Accessed several times 2002–2012. www.ipu.org/parline-e/reports/2045_E.htm.

Karatnycky, Adrian, Alexander Motyl, and Boris Shor. *Nations in Transit 1997: Civil Society, Democracy and Markets in East Central Europe and the Newly Independent States*. New Brunswick: Transaction Publications, 1998.

Karatnycky, Adrian, Alexander Motyl, and Aili Piano. *Nations in Transit 1999–2000: Civil Society, Democracy and Markets in East Central Europe and the Newly Independent States*. New Brunswick: Transaction Publications, 2000.

Karatnycky, Adrian, Alexander Motyl, and Amanda Schnetzer. *Nations in Transit 2001: Civil Society, Democracy and Markets in East Central Europe and the Newly Independent States*. New Brunswick: Transaction Publications, 2002.

——. *Nations in Transit 2002: Civil Society, Democracy and Markets in East Central Europe and the Newly Independent States*. New Brunswick: Transaction Publications, 2003.

Kolarova, Rumyana. "Democratization in Bulgaria: Recent Trends." In *Democratization in Central and Eastern Europe*, edited by Mary Kaldor and Ivan Vejvoda, chapter 11. New York: Pinter, 1999.

Nagle, John and Alison Mahr. *Democracy and Democratization: Post-communist Europe in Comparative Perspective*. London: SAGE Publications, 1999.

Nucleonics Week. January 1, 1993: 1; February 4, 1993: 1; May 13, 1993: 1; June 24, 1993: 1, 13; December 9, 1993: 10; September 21, 1995: 1; October 12, 1995: 1; September 12, 1996: 5; December 11, 1997: 1; April 23, 1998: 7; March 18, 1999: 13; September 16, 1999: 12; October 21, 1999: 18; June 22, 2000: 6; January 31, 2002: 1; July 4, 2002: 1; July 11, 2002: 1; August 15, 2002: 1; October 3, 2002: 1.

Taras, Ray. "The Politics of Leadership." In *Developments in Central and East European Politics, Volume 2*, edited by Stephen White, Judy Batt, and Paul Lewis, chapter 6. Durham: Duke University Press, 1998.

Western European Nuclear Regulators Association. "Nuclear Safety in EU Candidate Countries," October 2000. Accessed multiple times 2000–2012. http://europa.eu.int/comm/dg1a/nss/index.htm.

Additional Reading

BBSS Gallup International. *Attitude towards Bulgaria's Membership in the EU, January 2000–May 2002*. 2002a.

——. *Perceptions of the Effect of EU Integration for Bulgaria, May 2002*. 2002b.

——. *Perceptions of the Importance of EU Membership for Bulgaria, May 2002*. 2002c.

Drezov, Kyril. "Bulgaria and Macedonia: Voluntary Dependence on External Actors." In *Democratic Consolidation in Eastern Europe: International and Transnational Factors, Volume 2*, edited by Jan Zielonka and Alex Pravda, chapter 15. New York: Oxford University Press, 2001.

Grant Agreement between the Republic of Bulgaria and the European Bank for Reconstruction and Development, Kozloduy Nuclear Safety Project. 1993.

Karatnycky, Adrian, Alexander Motyl, and Charles Graybow. *Nations in Transit 1998: Civil Society, Democracy and Markets in East Central Europe and the Newly Independent States*. New Brunswick: Transaction Publications, 1999.

Official Journal of the European Commission. *Bulgaria: Accession Partnership*. C202/1, June 29, 1998.

——. *Written Questions to the Commission and Replies on Behalf of the Commission*. 98C 174/92–C40 E/42, November 19, 1997–July 27, 2001.

——. *Council Decision of January 28, 2002 on the Principles, Priorities, Intermediate Objectives and Conditions Contained in the Accession Partnership with Bulgaria*. 2002/83/EC, February 14, 2002.

6 Overcoming Slovakia's Meciar problem

"The problem with Meciar was that he behaved arrogantly with people. He couldn't play the game. That was stupid."

—former senior Slovak government official, September 2002[1]

"After Meciar, EU membership was the first priority for the Slovak government. I was only given ten minutes to present on the safety of Bohunice, and I had to fight to even get invited to the meeting with the EU. It was a political decision not based on safety."

—Miroslav Lipar, former Chairman Slovak Nuclear Regulatory Authority (UJD) (1997–2001), September 2002

Introduction

The fall of 1999 brought sighs of relief across Western Europe. In late September, Slovak officials agreed to close two of the country's oldest and most unsafe reactors by 2006 and 2008, two years earlier than their respective thirty-year design lifetimes. The decision came after the state utility had invested several hundred million dollars to enable the plants to run at least until the early 2010s. The commitment to shut down the reactors represented a previously unattained level of compliance with a 1993 commitment to close the V1 complex at Bohunice.

A few months before the landmark decision, Vladimir Meciar had been soundly defeated in the May 1999 presidential elections. Meciar had acted as the catalyst for Slovakia's political instability ever since the "Velvet Divorce" from the Czech Republic in 1993. Even though his party, the Movement for a Democratic Slovakia (HZDS) had won the 1998 parliamentary election, it could not attract enough coalition partners to form a ruling government. When opposition parties took over later that year, the country finally began a steady movement away from the previous five years of intense political instability.

Those early years of independence had been characterized by formal instability between the various branches of government and arbitrariness in the informal practice of politics. However, while the formal rules stabilized, stable governance took much longer to appear. The process reached a critical juncture after opposition candidate Rudolf Schuster defeated Meciar in the May 1999 presidential race, thereby significantly reducing the threat of Meciar's return to power. Yet Meciar did not own all the blame for the country's problems; the rules themselves were unstable and compliance suffered as a result. Then as the rules stabilized, compliance behavior improved as officials became more willing to accept considerable short-term costs in order to gain the long-term benefits of cooperation with the West.

Compliance began the same year the Velvet Divorce divided the former Czechoslovakia as the new Slovak government and the European Bank for Reconstruction and Development (EBRD) agreed in 1993 to nearly $600 million in loans and credits to fund completion of two nuclear reactors at the Mochovce plant. In return, Slovakia committed to close its oldest reactors at Bohunice within the next few years. Along with signature of the Convention on Nuclear Safety (CNS) in 1994, this commitment became the basis for measurements of Slovak compliance over the ensuing decade. A similar pledge to shutdown two more first-generation reactors at Bohunice given to the European Commission several years later secured the country's entry into the EU candidate group.

Slovakia originally had four operational VVER units at Bohunice, two 440-230s (labeled V1) and two 213s (V2) similar to the two newer reactors at Mochovce. Despite smaller pipe rupture figures in the event of loss-of-coolant event, the 213s were actually somewhat safer than the 230s given their higher quality components and safety redundancies. Despite the presence of a bubbler condenser tower to act as a confinement system that reduced the potential release of radiation in a catastrophic accident, V1 still had so many structural problems that it could never reach comparable Western safety levels.

When the Soviet Union and Slovakia first planned to build a nuclear power plant at Bohunice in 1978, the Russian team decided that the 213 model was too expensive and convinced the Slovaks to build the cheaper, less safe 230 reactor. From the beginning of construction though, the Slovak ministry of atomic energy and nuclear plant engineers endeavored to compensate for the plant's design weaknesses. Former nuclear regulatory chairmen Jozef Misak and Miroslav Lipar expressed pride that V1's reactors were not like other VVER 230s due to the extensive safety modifications built into the plants throughout their operating lives.[2] They explained that plant management immediately began safety upgrades after starting up Bohunice's first reactor. Then after the Chernobyl disaster, the Soviet Ministry of Atomic Energy ordered improvements at Bohunice and work continued through 1989, after which Slovakia initiated its own Small Reconstruction in 1991, followed by the Major Reconstruction from 1995 to 2000. Yet despite the improvements, V1 fell into the category of first-generation Soviet reactors that could not attain comparable safety standards to Western reactors. The G7 pushed for early closure of V1 from as early as 1992, and argued that V2 needed extensive, expensive upgrades that could delay its early closure, but Western officials had serious doubts Slovakia could pay for them.

Like other nuclear power plants in Eastern Europe and the former Soviet Union, Bohunice got caught between competing concepts of safety in those discussions. While the IAEA consistently gave the reactors favorable safety ratings,[3] as did some Western experts after the Major Reconstruction achieved marked success in upgrading certain safety features,[4] Western governments maintained the position stated in 1992 that the reactors could never reach an absolute level of safety because they lack containment structures.[5] In particular, V1 got signaled out due to serious doubts about the strength of the reactors' pressure valves, which could burst with the sudden inflow of cold water in the event of a core accident. However, Slovakia's international agreements recognized the differences between V1, V2, and the newer Mochovce reactors, and really only pushed for V1's early closure. The others required upgrades, but given their safer components, and the later addition of a confinement system similar to V2's, the newer reactors were exempt from early closure requirements.[6]

The distinction had little immediate impact as Slovak compliance with the agreement requirements remained inconsistent throughout much of the past decade as competition between rival politicians challenged the formal rules of the game and destabilized the informal rules with arbitrary governance procedures. In addition, Slovakia had retained a penchant for a "strongman" leader longer than other post-communist countries in Eastern Europe, one determined to rule the system regardless of precedent or prescribed patterns of governance. This figure also engendered intense rivalries bent on stopping him, sometimes regardless of the cost. Vladimir Meciar was that catalyst in Slovakia's first decade of independence.

Bright beginnings gone bad

From the earliest days of *glasnost* and *perestroika*, Vladimir Meciar dominated Slovakia's political scene. He used his position as chairman of the Movement for a Democratic Slovakia (HZDS) to push forward the independence movement even when the general public remained more dubious of its benefits.[7] In the 1992 elections leading up to the breakup of Czechoslovakia, HZDS gained a plurality of the vote largely due to Meciar's assurances that the transition to a market economy would take into consideration the vital role of state involvement in the economy. As an example of this, once elected, Meciar stopped privatization in its tracks despite considerable progress toward marketizing the Slovak economy. After his first brief ouster from power in March 1994, the subsequent government reintroduced coupon privatization, which gave citizens shares in formerly state-controlled companies, but upon returning to power, Prime Minister Meciar invalidated all sales, even those to foreign investors.[8] Considerable effort and costs had gone into those programs and Meciar arbitrarily disregarded them in what would become a hallmark of his time in office. In addition, his personal appeal to average working-class Slovaks as their protector against ruthless "marketizers" created an individualistic policy-making style that often conflicted with nascent democratic rules and institutions. That kind of personalized politics added fuel to the growing conflict between competing elites for the course of Slovakia's political development.

To make matters worse, Meciar had his hands in everything, including nuclear safety. He was not the only one trying to control the nuclear industry, but his position as Prime Minister gave him considerable power over political rivals, although his opponents regularly tested the limits of that power. One former senior Slovak government official told me that while on the one hand Meciar supported efforts to improve nuclear safety, he also refused to allow independent action by the regulatory agency, one of the key stipulations in Slovakia's international agreements. The official complained that he "had to take a lot of preparations with documents because [Meciar] would study them. But he wouldn't accept any discussion at times. He was just issuing his decisions."[9]

In spite of these problems in government, Slovakia received a significant aid package from the EBRD and the European Commission (EC) totaling nearly $600 million, and while the money did not go directly into the state treasury, Slovak companies competed for and completed safety work at Bohunice. Yet the agreements contained considerable requirements, all of which had explicitly spelled out the West's commitment to fund completion of Mochovce 1 and 2 on the expressed condition that V1 close earlier than its official thirty-year design lifetime.[10] Unfortunately, as early as September 1994 rumblings began spreading west that Slovakia was considering a major upgrade at V1. One Bank official identified the coming problems by arguing that after reconstructions costing $200 million, "no one will be convinced that [the Slovaks] will shut it down in 1999 or 2000, and nobody on the EBRD board would believe us if we said so."[11] The problem of Meciar's arbitrary decision making in other policies had carried over to Bohunice when he changed the safety work from facility upgrades and overhauling operational procedures intended to ameliorate some of the fundamental structural problems, to efforts at making V1 viable in the long term as a way to get out of the original 1993 early closure commitment.

The Slovak constitution allowed Meciar to make those decisions, but the formal rules of the game had problems from the start. Even though the document created a legislative system, it also established an unclear division of political authority, thus creating the preconditions for unstable governments in general.[12] Making matters worse, those formal rules contrasted the informal rules since 1994 when Meciar took control of governance and thereby threatened the formal rules through what the Deputy Prime Minister later called "unusual decision making."[13]

Opposing formal and informal rules

The constitution gave nearly total control to the political majority in Parliament as the government formed from that majority largely determined cabinet positions and controlled the legislative process. In contrast, parliamentary opposition and the President had very limited influence.[14] Yet throughout Meciar's time in office and the time leading up to his defeat in the 1999 presidential election, the rules of the game were repeatedly challenged by the Prime Minister, his cabinet, opposition groups within Parliament, and his main rival, President Michal Kovac. Whenever Prime Minister Meciar proposed alternative institutional structures to enhance his personal power, his opponents escalated their attempts to constrain him in return through legal and other means. Consequently, uncertainty in the political system went deeper than normal conflicts over various policy programs, instead centering on the very nature of the rules themselves.

Given the Prime Minister's authority as the legal head of government, Meciar attempted to become the unrivaled ruler of the country. The President still retained a central role in governing the country, but the position had severe limitations imposed by the constitutional powers of the Prime Minister and the cabinet around him. Rather than roll over, President Kovac became the proverbial thorn in Meciar's side. While he originally ran on the HZDS ticket with Meciar's initial support, Kovac soon proved himself fiercely resilient to the Prime Minister's regular attacks on the institution of the presidency. Those efforts did not solve the fundamental problem of the system though since Kovac also operated under generally uncertain constitutional provisions that formally codified a separation of governing powers, but failed to establish clear mechanisms for resolving disputes between the Prime Minister and the President. Yet while the constitution had structural problems, the politicians themselves contributed greatly to the underlying instability of the time.

The conflict between President Kovac and Prime Minister Meciar began soon after independence. As early as 1993, Meciar began to remove any officials who stood against him, whether initially loyal to him or not, despite constitutional limitations on his ability to do so. In what would become a long-standing practice, President Kovac regularly petitioned the Constitutional Court to reverse and stop Meciar's firings. The Court ruled that only the President has dismissal authority, whereas the Prime Minister can simply propose such action,[15] yet the constitution did not offer any mechanism for enforcing the Court's decisions and Meciar continued unhindered. The problem stemmed from the unclear constitutional provisions delineating the President's power from that of the government, thereby enabling Meciar to pursue absolute power opportunistically. However, Meciar's greatest effect was not to destabilize the system by his actions alone. More troublingly, his opponents also challenged the Prime Minister's legal authority, sometimes with unconstitutional and arbitrary legislation, all of which helped to maintain a significant level of uncertainty within the political system.

In an effort to resolve the situation by removing the troublemaker, Kovac went after Meciar in March 1994 by calling for a vote of no-confidence, which was not one of his legal powers. The tactic worked in the short term but did not remove Meciar from political contention. Despite his ouster, HZDS won a considerable victory in the ensuing parliamentary elections six months later, and two months after that Meciar resumed his position. While this showed that the electorate still supported Meciar in significant numbers, the effects remained the same in that he could continue as a lightning rod for destabilizing the political system. However, the second time around Meciar lacked a legislative majority to pass amendments designed to strip the President of his legal authority and power,[16] even though he repeatedly attempted this through extra-legal means. Stymied by the formal provision, Meciar called for a no-confidence vote in Kovac in May 1995, although it possessed no legally binding force, a fact the Constitutional Court affirmed the following November. In response, Meciar savagely cut the President's budget and tried to strip him of key appointment powers, first by taking away his ability to appoint the head of the Slovak Intelligence Service (SIS), then by transferring the

President's power to designate the army chief of staff to the cabinet. The Constitutional Court ruled in November 1996 that while the first move found support in the constitution, the second did not. Even this decision did not perfectly line up with the constitution since article 102 gave the President the power to appoint and dismiss high-ranking officials, with the exception of the Prime Minister, which clearly includes the SIS director.[17]

Even more troubling proved the association of HZDS government officials with politically motivated violence beginning after the 1994 parliamentary elections.[18] The clearest example involved the kidnapping of President Kovac's son. The timing and style of the act, as well as the reactions of key representatives during the course of the investigation indicated the underlying political motivations.[19] SIS chief Ivan Lexa stood out as the ringleader, likely under Meciar's personal direction given his control over party patronage and government decision making. According to official reports, unknown assailants abducted the boy, got him drunk, and took him to Austria. While some may consider that a typical college vacation, the President was not amused and the young man suffered considerable injuries in the process. I do not mean to marginalize the violence of the incident or the trauma caused by the abduction. My point is simply to state that while Meciar may have crossed some lines, he remained within important boundaries since the young man did not suffer serious harm or worse. Therefore, the incident indicates more of the Prime Minister's destabilizing practices than a murderous level of political violence.

Although Meciar did not confine his attempts to trying to usurp presidential authority, he seemed to take particular delight in it. In one example, he ordered a huge clock constructed within clear sight of President Kovac's office. The clock did not actually keep time, but rather ran backward counting down the days of Kovac's term. Even this was done arbitrarily though. The time did not mark the days left as determined by the constitution, which measured the time from when the President took office; Meciar had it set to count down from the President's election date. Opposition deputies repeatedly asked the Prime Minister to relent, but he refused to change the time. In contradiction to the law, Meciar insisted that Kovac step down on the earlier date, "even though the difference in time was trivial and made no difference whatsoever in substantive political terms. The clock served as a constant reminder to Kovac and all of Bratislava that Meciar's whims soared above the law."[20]

The Prime Minister's growing megalomania also made the processes of governance unstable. Meciar and his HZDS followers frequently disrupted even the most rudimentary parliamentary procedures. The problems started early after the October 1994 elections when Meciar tried and failed to remove the election mandates of fifteen Democratic Union (DU) officials after they switched party platforms on instructions from President Kovac. If successful, this would have given HZDS a three-fifths constitutional majority and enough power to realign the constitutional framework to Meciar's advantage. HZDS petitioned the Constitutional Court to support its claim but the Court eventually denied the request. Meciar responded by ordering Interior Minister Ludovit Hudek to investigate every signatory on DU's original petition to participate in the election. This too failed when over ten thousand people came forward and confirmed their support for the opposition party.[21]

When these overt attacks on his political opponents ultimately failed, he tried to introduce a draconian form of party discipline on HZDS by punishing dissenters if they left the party. Election laws did not clearly require strong adherence to the party line, so informally Meciar tried to strengthen obedience by forcing elected members to sign a preemptory letter of resignation in case they ever voted against the party, or more precisely against Meciar. Frantisek Gaulieder did just that in December 1996, but his gamble did not pay off as the party leadership simply filled in the date on his resignation letter and sent it to the parliamentary Immunity and Mandate Committee. Gaulieder tried to forestall the inevitable and sent letters saying he had no intention of resigning. His house was subsequently bombed.[22] As was the case in so many instances in Slovak politics after 1993, rival groups called upon the Constitutional Court to help define the rules of the game, this time the processes by which ruling parties resolved internal dissent. As was equally common though, the Court's decisions in July 1997 did very little to resolve the uncertainty and arbitrariness of governance under Meciar.[23] The Court agreed with Gaulieder saying that Parliament (controlled by HZDS) had violated his constitutional rights, but he could not return to duty until the legislature reversed its decision since the Court did not have that kind of authority. Several times opposition parties tried to resolve the issue but each time the ruling parties vetoed or simply boycotted the proceedings by walking out of Parliament. This unexpected quorum busting destabilized the informal rules further by making governance even more difficult.

The pattern continued when Emil Spisak was denied his legitimate place in Parliament after Bartolomej Kunc died and his party refused to give Spisak, the next name on the party list, the spot since he had allegedly betrayed the party earlier in his career. The Constitutional Court ruled that Spisak's rights had been violated but as with the Gaulieder case, it could not compel Parliament to give him the seat since only Parliament can retract its own decisions. The Gaulieder and Spisak cases showed the extent to which legislative appointments and delegations of power between and within political parties were regularly contested, and how the court system only vaguely resolved the issues.

Meciar costs the country

Systemic uncertainty carried over to Slovakia's international relations with the West as well. In 1995, the government decided to forgo early decommissioning of V1 based on domestic energy needs when in March of that year Meciar stated that he would not shut down V1 without a clear "guarantee that Mochovce 1–2 would be commissioned by then," adding that the EBRD's insistence that he raise electricity rates by twenty-five percent "would bankrupt most small businesses in the country."[24] Chairman of the French contracting firm EDF, Gilles Menage, met with the Slovak Prime Minister around the same time and said Meciar became incensed by the West's insistence that Slovakia comply with its commitment to shut V1 as a condition of assistance, and responded "I have decided to finish Mochovce, and I'm asking for resources to do it, not for permission."[25]

Things began to unravel shortly thereafter as Bayernwerk AG and EDF, the principal Western contracting companies working on Mochovce, pulled out due to the unfavorable political situation.[26] Meciar's government responded by proceeding with plans to upgrade V1 even though it drew harsh criticism from the West. According to state utility senior manager Karol Bodorik, "The EBRD had no basis for expecting us to shut V1. V1 provides Slovakia a quarter of our electricity. The plants are safe. We are going ahead with this, whether they like it or not."[27] Short-term interests mattered more than long-term planning for Slovak decision makers, forcing the European Union to use its trump card by voting to pass over Slovakia in the first-round of membership negotiations.

Slovakia's international agreements spelled out the costs of non-compliance, and the European Union compounded those costs with the threats to withhold future cooperation. Yet despite being sidelined, elites could not see the loss of the benefits clearly due to the rampant instability in the rules of the game. Challenges to the constitutional structures and arbitrariness in government decision making by Meciar and his opponents created uncertainty about long-term planning, so short-term needs and the high costs of compliance overshadowed the potential future benefits of EU membership. However, it is important to note that opposition and governing parties had a general agreement about Bohunice's future, that all of its reactors should operate at least as long as their design lives. The convergence indicates that compliance problems did not simply fall prey to policy differences. The problem stemmed from much larger ongoing destabilization of already unstable political rules. Meciar and his supporters, along with those who fought against him, added to the uncertainty of the times so that even when officials just tried to do their jobs, Meciar would often supplant, ridicule, and generally limit their ability to function properly. Those opposed to him did the same thing, helping to create an abiding and pervasive uncertainty as to how things would get done and by whom.

Coupled with unstable informal rules, uncertainty about the formal mechanisms of power transfer, unclear differentiation of delegated authority between the government and the President, and disagreement about those divisions made medium and high compliance unlikely. Low compliance carried definite costs as well, but the Slovaks had been working toward increasing operational safety all along so low compliance did not require as much of a leap as other levels. In particular, uncertainty about ministerial responsibilities as part of Meciar's arbitrary style of governance prevented the establishment of a sufficient nuclear regulator, the condition for medium compliance. Early closure was also clearly out of the picture since keeping V1 running met many short-term needs, and short-term thinking ruled the day as politicians could not realistically plan much beyond the immediate future. Therefore, the need to defend their present interests framed medium and high compliance commitments as too costly.

Formal election laws and informal party systems played a central role in defining those interests, specifically future prospects for remaining in power. These remained in flux during Meciar's reign as HZDS and its coalition partners attempted to change the rules in order to improve their chances of remaining in power.[28] Under the Prime Minister's direction, Parliament voted in March 1996 to redistrict the entire country to form a single administrative unit. This gave Meciar two powerful advantages. It enabled charismatic leaders to capitalize on their support throughout the country rather than have it squandered in a few localities, and it also gave more power to the government-controlled Ministry of the Interior by reducing the Central Electoral Commission's ability to ensure free and fair elections. The change was not based on a detailed study of public opinion, options for more efficient election procedures, or even consensus with opponents in Parliament. Meciar used his considerable power to shape the political system to his personal advantage, which kept politicians guessing about the actual processes by which the country would be governed. Nearly two years later, in May 1998 Meciar pushed through another law that forced smaller parties to coalesce into larger single parties or be forbidden to compete in the general elections. He seems to have hoped this would further weaken the opposition. While the tactic did not produce the election results Meciar wanted in the long term, it led to an even greater sense of instability in the rules of the game as party officials scrambled to redefine their platforms and the electorate struggled to keep up with the changes.

The extent of Meciar

It would be a mistake to assign blame to a single person or group for Slovakia's political instability. The problems ran throughout the political system, and reflected a lack of consensus about fundamental views of democracy and the rules governing the country. Making matters worse, very little substantive dialogue existed between opposing groups to try to resolve their common problems. Yet while not singularly responsible, Meciar acted as the catalyst for instability as others responded to his attacks and "fought fire with fire" to minimize his legitimate power. He approached them confrontationally with arbitrary and quasi-authoritarian tactics to consolidate patrimonial power under the conditions of what he hoped would become a façade of democracy without the substance of democratic rules.[29] His behavior extended to his own party, preventing HZDS from building and maintaining broad-based, lasting political consensus, all of which eventually brought him into conflict with some of his former allies.

Throughout his time in power, "Slovakia was governed by a coalition of misanthropes and harlequins, headed by a politically wily but mentally unbalanced Prime Minister [who] established a regime based on thuggery, incompetence and contempt for the law."[30] In particular, Meciar arbitrarily replaced competent bureaucratic and party officials with unqualified lackeys, and anyone with "triple-digit IQ's" was a threat."[31] When opposition members questioned these loyal ministers during open parliamentary debates, their ineptitude was made painfully obvious by the Prime Minister himself.

> He would rise from his seat, elbow the befuddled minister under questioning away from the podium, and declare that he would tackle the query himself, demonstrating that only he was capable of shouldering the weighty task. This reduction of even top government officials to a cadre of hapless lapdogs suited Meciar perfectly, underscoring his indispensability and superiority.[32]

Frantisek Sebej, a leading liberal intellectual, noted that even the communists had rules they followed; not so with Meciar.[33] "Meciar reveled in arbitrariness and in breaking even those laws that he had made, or earlier acquiesced to himself. There was an almost anarchical abandon in the manner in which he flouted and undermined established procedures."[34]

A clear example of this occurred in December 1996 when opposition parties sponsored a proposal to hold a public referendum on direct presidential elections. At the time, the constitution required Parliament to select the President with a three-fifths majority.[35] Meciar refused to allow the referendum and pointed out that the constitution did not specifically allow for popular plebiscites to change presidential appointment procedures. The opposition countered that the constitution locates the real power to govern with the citizenry, thus making referenda a legal form of change.[36] Both appealed to the Constitutional Court, which ruled in May 1997 that the referendum was legal but the wording of the ballot question was not. The government used the ruling as a vague justification for removing the question on direct presidential elections from ballots, leaving only three questions about joining NATO.[37] Low turnout and a high proportion of uncompleted ballots turned in by voters refusing to sign the doctored questions invalidated the results.

Parliamentary opposition groups and the President then countered Meciar and maintained that the Court had instead legitimated their argument that the referendum itself was legal. President Kovac called for a new referendum in April, but Meciar effectively canceled this by assuming some of Kovac's presidential powers. "This decision further intensified institutional and political conflicts, as it was unclear whether he had the authority to do so."[38] The fact that both sides continued to use the Court as both an arbiter for political conflict, as well as the final constitutional authority, could have been a stabilizing factor in Slovakia. Regrettably, the fact that both sides interpreted the rulings so differently shows the rules defining the method of power transfer were unpredictable and open for challenge, despite the Court's efforts.[39]

These problems continued throughout 1998 as Parliament tried and failed to elect a new President after Kovac's term ended.[40] For six months neither side could muster the necessary three-fifths majority to approve a candidate, and after the fifth attempt failed, Meciar simply took over the job. In a symbolic act to show his dominion over political life in the country, Meciar announced that his cabinet would meet in the presidential offices from that day forward.[41] While all the preceding parliamentary efforts had taken place within the structural context of the constitution, the actual process of determining who would govern from the President's office was thrown off balance due to Meciar's illegal maneuvering.

Compliance suffers

Instability in the rules of the game and low compliance under Meciar did not just affect politics or even Bohunice V1; the same systemic uncertainty affected the more modern and safer Mochovce as well. The new plant became operational before workers completed all the required safety modifications, as Western companies working on the projected cited "intense pressure from the Slovak leadership to get Mochovce 1 online."[42] Even if the regulatory authority had wanted to slow the pace, it had little autonomy with Meciar in power, something Slovak officials had committed to rectify when they signed the Convention on Nuclear Safety in 1994. Recalling his tenure under Meciar, Miroslav Lipar said that 1997 and 1998 was a "very challenging time."[43]

The regulatory authority officially had entered the fray a few years earlier in 1993, and once Lipar took over from Jozef Misak in 1997 the organizational staff increased. However, it still lacked the autonomy and authority necessary to ensure compliance with the CNS,[44] although this lack of autonomy did not produce any substantive policy discord between the Prime Minister and nuclear industry officials. Both former regulators told me that they supported Meciar's plans to continue operating Bohunice despite international commitments to shut it down. Almost immediately after the government entered into the EBRD Grant Agreement, Misak had refuted the need to decommission V1 in order to receive assistance at Mochovce, saying that the nuclear regulatory authority "would accept no connection between Bohunice closure and Mochovce completion,"[45] even though the EBRD agreement made that connection explicitly clear. A former Slovak official agreed, saying that he "couldn't understand that the [1993 government] statement was made without technical evaluations. For me it was certainly not fair."[46] Lipar confirmed the view that "V1 could operate much longer than the shutdown agreement of 2000."[47]

The EBRD agreement stipulated that the regulator ensure compliance with safety upgrades and decommissioning efforts as prescribed by the bank, not their own evaluations. Despite personal agreement with Meciar's decisions, the pattern of insufficient regulatory autonomy to fulfill agreement requirements coincided with short-term thinking throughout the political elite. Autonomous power was not something Meciar was willing to grant the regulator, even though those working for him agreed with his decisions about Bohunice. The possibility that someone would challenge his authority from within the government structure had always been a serious concern for Meciar, and since he could not take that chance and give up control, he kept friends and foes guessing what his next move would be.

The end comes none too quickly

Prior to the 1998 parliamentary elections, Meciar tried to use his previously usurped presidential power to grant Ivan Lexa immunity from prosecution for his reported involvement in the 1995 kidnapping of Kovac's son. He attempted the same for Gustav Krajci, the former Interior Minister who single-handedly destroyed the previous referendum on direct presidential elections.[48] Yet once again Meciar did not act alone in creating these problems. After his party failed to form a governing coalition after barely winning the elections, the new opposition-led Slovak Democratic Coalition (SDK) retaliated by revoking Krajci's immunity despite the Constitutional Court stating this violated the constitution. Showing that instability extended beyond one individual to impact the political system generally, "officially nobody from the government or the Parliament declared that the Court's decisions should not be obeyed . . . The government [instead] delayed the implementation of or simply ignored the Court's rulings."[49]Anti-Meciar groups also tried to take away one of Meciar's legislative protections when the newly installed government fined him for failing to answer certain questions about Lexa and Krajci. Meciar petulantly responded that he would go on a hunger strike if Parliament tried to take away his oath of silence. The law considered the oath inviolable, but that fact did not stop Meciar's enemies from threatening the former Prime Minister. Therefore, at best both sides continued to appeal to the same set of laws, but with diametrically opposed meanings attached to their pronouncements. At worst, they simply ignored them to get at their rivals.[50]

Despite these problems, the September 1998 parliamentary elections initiated a new period in Slovakia's political system characterized by potential movement toward either increasing stability or a return to the past. The aforementioned examples reveal the durability of bad habits among the country's political elites, but Mikulas Dzurinda's government attempted to bring Slovak politics out of its vicious cycle. The breathing room it had after gaining a tenuous hold on power afforded enough opportunity to try new things but not enough to take great risks. As a result, the new government began to improve the status of the nuclear regulator almost immediately. Prime Minister Dzurinda increased the payroll of the Nuclear Regulatory Authority (UJD) and gave it more autonomy to determine the safe operation of Bohunice and Mochovce. The Western European Nuclear Regulators Association (WENRA) and the Second Review Meeting of the CNS Contracting Parties confirmed that the Slovak regulator's new "status [was] comparable to that of the regulatory bodies in Western European countries."[51] These actions enabled Slovakia to meet the conditions for medium compliance due to increased stability in the rules of the game without Meciar wrecking havoc.

However, the rules did not fully stabilize after Meciar's ouster from power in 1998. From September 1998 until May 1999 Meciar still retained the possibility of being a catalyst for renewed instability. Undaunted by his earlier defeat, Meciar began his run for the presidency shortly after HZDS lost the top spot in Parliament. Those intervening nine months before the presidential election did not reach the level of high stability because of the very real threat Meciar would pose if elected President. The Dzurinda government had come to power because HZDS could not attract enough smaller parties into its coalition. The new government's position was anything but secure though. Its grasp on power depended on the continued goodwill of its coalition members, but more importantly the support of the people. For that reason, the change in government did not produce high compliance because the costs were so enormous and most of the Slovak electorate opposed early closure. Officials had greater incentive to honor their commitments for bolstering the nuclear regulator's autonomy, but the threat of Meciar's return still loomed large on the horizon and closing V1 at that time would cost too much politically.

Consequently, the new government did not immediately change policies regarding V1 shutdown despite improvements to the regulator's autonomy at the same time. In compliance with the CNS, the government deferred to Lipar and his regulatory staff to determine the appropriate course of action, and following their recommendation in April 1999 officially renounced the earlier commitment to shut V1 by 2000, even though it meant partially breaking Slovakia's commitments to the EBRD and the EC. While the regulatory agency had sufficient autonomy and authority to make and enforce its decisions for the first time, thereby ensuring compliance with the CNS, the government's overall compliance behavior would not move any higher until Meciar no longer threatened the system.

Had Meciar succeeded in any of his attempts to return to power, he could have easily reintroduced serious changes to government procedures and structural rules. That latent destabilizing effect diminished over time as Meciar was reduced to a vocal opponent who continued his ranting whenever anyone would listen in Parliament. Thankfully, he had little real power to challenge the rules of the game as an ordinary MP, even though he retained influence with popular opinion, but HZDS would not be able to call for early elections as long as the Dzurinda coalition remained cohesive, and non-binding referenda would merely signal public protest rather than change the current configuration of political power.

Only after the immediate threat of Meciar's return to power had receded through his resounding defeat in the second round of the 1999 presidential elections (forty-three to fifty-seven percent in support of the SDK supported Rudolf Schuster), and the new government had a realistic opportunity to work effectively without his presence to disrupt the rules and incite others to disrupt them, did the new government signal its intent to take the costliest step of decommissioning Bohunice V1. Prior to May 1999, the opposition government could not risk undertaking a decision that would add political clout to HZDS's claims that it intended to sacrifice Slovak interests for EU membership. The decision also went against the nuclear regulator's protestations that the reactors had reached comparable safety to Western plants, and therefore should continue operating for the remainder of their design lives. Yet elites had finally made the hard decision because they could finally see and appreciate the long-term benefits of the short-term costs of high compliance.

The Dzurinda government maintained a consistent pro-EU position, and the delay in moving from medium to high compliance did not result from a policy shift, nor was it due to a weakening of the bargaining criteria imposed by the European Union. Instead, the delay came from the need to be certain, or at least sufficiently certain given the structural limitations imposed on the parliamentary opposition, that Meciar could not throw the political game into turmoil again. Long-term thinking did not begin right away; it developed over time as the rules stabilized and the Dzurinda government came to accept that high short-term costs were worth paying to gain the long-term benefits of being in the European Union.[52] Those changes enabled the new Slovak government to recommit itself to early shutdown of V1 by 2006 and 2008,[53] establish a decommissioning fund to pay for its eventual closure, and develop a new energy strategy to compensate for the loss in generating capacity.[54] As these three actions met the required conditions for high compliance, the European Union invited Slovakia to join the club, and V1 closed on time.

Conclusion

The rules of the game break down into three periods of increasing stability that corresponded to increasing compliance. The first period of low rules stability and low compliance occurred from the beginning of independence in 1993 until HZDS's defeat in the September 1998 parliamentary elections. Formal and informal rules were uncertain, and Slovakia's nuclear community completed little work beyond safety improvements. The second period of medium stability and medium compliance took place as pro-constitution/anti-Meciar officials gained control of the political system but still faced the realistic threat of his arbitrary rule leading up to the May 1999 presidential election, due in large part to the constitutional revision in January allowing for a popular election of the presidency. During this period, any reasonable political actor would have assumed that Meciar would use the presidency as a base to continue his assault on the formal structures of government, and in the process destabilize the informal governance procedures as well. If Meciar had won the presidency, he would have likely used his popular mandate to usurp legislative and governmental powers, as he had done as Prime Minister. No institution seemed sacrosanct to him; all served as instruments to increase his personal power.

After taking office, the Dzurinda government resolved the structural problems in the formal rules of the game by clarifying election laws and the delegation of authority between government institutions, most notably dealing with presidential powers. Those changes grew more widely accepted since Meciar and HZDS could not counter them in Parliament. Informal rules remained more problematic since it took time for these new leaders to learn how to govern without Meciar as a demagogue to resist. New patterns of governance needed to become entrenched without Meciar's constant meddling or opposition efforts to limit him through unconstitutional means.[55] Meciar still had influence when it came to implementing government policies, so the day-to-day practice of politics remained uncertain. Uncertainty affected compliance since decisions to implement full closure commitments necessitated long-term thinking and medium rules stability made improving regulatory capabilities a less costly choice than full decommissioning. In contrast, the short-term costs were so great that officials needed to see a brighter future down the road before they could hope to make up for immediate losses.

That future also needed to look realistic, which only happened in the final period after Meciar's defeat in May 1999 removed the threat for several years and ushered in a new phase of high stability in the formal and informal rules of the game (the final time period for Slovak compliance). Widespread elite support from Dzurinda's broad-based coalition resolved unclear delegations of power and power transfer mechanisms within the state bureaucracy, and this allowed elites to count on continuity in the day-to-day practices of governing the country.[56] The threat had not completely disappeared as seen in HZDS's attempts to call for early elections in 2000 – without a legal basis and in contravention of the constitution – and in HZDS's election victory in September 2002, when it once again gained the most seats in Parliament but with an ever shrinking percentage of votes, a trend that would continue for the next decade until Meciar and HZDS did not even qualify for a seat in Parliament in 2010.

To that point, a good sign of the willingness to move beyond Meciar and his debilitating practices showed when a majority of parliamentary parties banded against HZDS rather than give Meciar another chance at power in 2002. High compliance became possible as elites gained breathing room to develop new long-term perspectives, and European integration itself offered an additional barrier to Meciar attempting to reintroduce instability in the rules of the game.[57] Western governments had little patience for his antics and resisted him through increased economic integration and Slovak dependence on the European Union.

In contrast to the model presented here, counter arguments based on policy priorities of different parties cannot account for the timing of Dzurinda's approaches to closing V1. High compliance did not occur when Meciar lost power in 1998; the government delayed until after his defeat in 1999. Vladimir Meciar was merely the catalyst for instability, not the sole cause of it as other political officials also operated under unstable rules, at times violating procedures and laws to achieve their agendas. Neither should Meciar's role appear minimal as he acted as more than just a thorn in the side of reasonable political actors. He destabilized the rules of the game, and until May 1999, the day-to-day procedures of governance remained uncertain due to the lingering threat of his arbitrary rule returning. Under such conditions, government officials willingly endured some of the costs of compliance, but given the tremendous short-term costs of closing Bohunice V1, the new Dzurinda government pursued medium compliance as a next step. High compliance resulted only after the threat of Meciar's return had substantively diminished.

Additionally, EU overtures to the new Slovak government do not explain the change in compliance because EU policy never changed; Slovakia needed to take the necessary steps for early shutdown to begin membership negotiations. The decision to pass over Slovakia in 1997 may have precipitated an electoral victory for democratic opposition parties, but it did not cause them to comply immediately and fully with Slovakia's nuclear safety commitments. The stability of the rules of the game varied, but EU policy did not.

Arguments based on safety culture fare just as poorly in explaining compliance. Slovak nuclear and political officials formed a tight-knit group. Plant personnel, regulators, and state utility managers generally came from the same academic institutions, possessed comparable knowledge and experience, and more importantly, held common views about safety culture with the country's political officials. That culture was generally risk-tolerant but not completely so. The emphasis on operational safety and the long history of safety upgrades from the very beginning of Slovakia's nuclear industry showed an appreciable aversion to risk as demonstrated by the introduction of a confinement system at Bohunice V2 and Mochovce 1 and 2, all of which met the European Union's minimum standards for safe operation and continue to operate today.

However, the same officials responsible for those upgrades also rejected the inherent necessity of a solid containment structure for their nuclear power plants indicating a more risk acceptant aspect to their safety culture than those espoused in the international agreements between Slovakia and the West. Containment structures prevent the release of dangerous radioactive material in the event everything else goes wrong. Confinement procedures do not comparably guarantee worst-case protection, and regrettably V2 still does not ensure absolute safety. A former senior Slovak government official repeatedly offered assurances that a major loss-of-coolant accident foreseen at Western plants would not happen at V2 or Mochovce. However, he admitted that if such an accident occurred, confinement systems could not protect the people in the surrounding regions.[58] Moderate risk tolerance in Slovak safety culture has therefore been a constant feature throughout the period of compliance. In contrast, compliance has changed despite resistance from safety officials in the nuclear regulatory agency and the politicians who supported them. Therefore, safety culture cannot explain variations in compliance. In contrast, rules stability and compliance progressed along similar trajectories at the same times. The EC noted this in several reports from August 1998 to September 2002.[59] In each case, Slovakia moved progressively toward EU accession through increasing stability in the rules of the game.

Table 6.1 Slovakia Time 1 (1993–September 1998)

Rules Stability	*Clarity*	*Predictability*
Formal	No	No
Informal	No	No
Conclusion: Constitutional formal rules left gaps that opportunistic elites exploited and the Constitutional Court did not completely resolve. Informal rules followed the elite competition between Meciar and his opponents; the Prime Minister catalyzed instability. The rules were neither clear nor predictable indicating **LOW STABILITY**.		
Compliance	• Safety Upgrades at V1 and V2 • Nuclear regulator did not possess necessary autonomy or authority • 1993 Grant Agreement defined compliance and early closure of V1, upgrades to V2, and continued operation only if least cost option	
Conclusion: Slovakia achieved **LOW COMPLIANCE**.		

Slovakia's journey over the past two decades has been a difficult one. After gaining independence, Vladimir Meciar infected the Slovak political system with debilitating instability in the political rules of the game, and his adversaries often used

Table 6.2 Slovakia Time 2 (September 1998–May 1999)

Rules Stability	*Clarity*	*Predictability*
Formal	Yes	No
Informal	Yes	No
Conclusion: The formal rules stabilized as the competition between Parliament and the President abated and the new government worked to clarify information rules regarding the functions and operation of government. The rules remained unpredictable as long as Meciar posed a threat to the system before the 1999 presidential election. This combination produced **MEDIUM STABILITY**.		
Compliance	• Safety work completed at V1 and V2 progressing on schedule • Nuclear regulator given autonomy and authority • Government rejected V1 early closure with regulators support	
Conclusion: Slovakia achieved **MEDIUM COMPLIANCE**.		

Table 6.3 Slovakia Time 3 (May 1999–2008)

Rules Stability	*Clarity*	*Predictability*
Formal	Yes	Yes
Informal	Yes	Yes
Conclusion: Formal and informal rules became increasingly predictable as Meciar's opportunities to regain power dwindled and elites operated in established patterns and according to prescribed formal rules. These indicate **HIGH STABILITY**.		
Compliance	• Safety upgrades completed • Regulator autonomous and funded • V1 decommissioned as required and closed on time.	
Conclusion: These actions brought the country into **HIGH COMPLIANCE**.		

similar tactics to challenge the boundaries of his power, further calling into question both the legitimacy of government structures, as well as the processes for governing the country. While the idiosyncrasies of the Slovak constitution placed a preponderance of power in the Prime Minister's hands, the lack of clear procedures for resolving conflicts between rival institutions only made matters worse. However, none of that would have been a problem without Meciar to stir things up and cause problems. He made a potentially challenging situation far more chaotic than it need have been. He bears much of the blame as he took advantage of the power granted him, manipulated the system in his favor, and continually tested the nerve of his opponents. Take out Meciar and Slovakia would have followed a different course. At the very least, the journey to close V1 and upgrade V2 would likely not have taken so long or caused so much anxiety about another nuclear disaster in Eastern Europe.

Notes

1 The official requested that I not mention his name. He may have felt free enough to make such a statement in private, given the improved political stability in Slovakia, but Vladimir Meciar remained a potentially dangerous figure in his country at that time.
2 Jozef Misak and Miroslav Lipar, in discussion with the author, September 2002.
3 Expert Mission of the IAEA Technical Cooperation Project (2000).
4 Notably Western European Nuclear Regulators Association (2000).
5 Before the Major Reconstruction Western European nuclear industry officials roundly condemned V1 (see *Nucleonics Week*, August 5, 1993: 1 for an example.) However, as the upgrade work progressed industry opinion fractured between French proponents and German opponents. Whether this had more to do with internal EU politics or different safety standards is not the central issue here, although it does factor into the discussion. The core of this project is Slovak compliance, not the basis of safety assessments per se.
6 Successive Austrian governments led the fight to keep Mochovce 1 and 2 from entering service and, despite eventually bowing to pressure from their EU counterparts, the Austrians remain ardent opponents to all Soviet reactors, especially those that ring their borders. Given the transnational contamination after Chernobyl, one cannot blame them for trying.
7 Zemko (2000): 254.
8 Wolchik (1997): 218.
9 Anonymous Slovak official, in discussion with the author, September 2002.
10 Miroslav Lipar explained his position, one supported by the IAEA, that design lifetimes are a subjective number based on the number of times a reactor trips. It was assumed by Soviet engineers that this would happen roughly once a year, and the reactors were designed to experience thirty trips. Lipar asked me, "What if the reactor only trips once in five years, or ten years? The issue is lifetime management not extension." He has a point, but it still contradicts the fundamental agreement requirement to shut down reactors that have no solid structure confinement. Operational safety and confinement systems can reduce the possibility of an accident, but without a containment structure the release of radioactive material would be considerable if those measures fail. A former senior Slovak government official confirmed this with me.
11 *Nucleonics Week*, September 8, 1994: 13.
12 Malova (2001): 370.
13 Ibid., 374.
14 Initial presidential powers included the ability to call for early elections and appoint high-ranking national security officials. These became the battleground between President Kovac and Prime Minister Meciar during their terms in office.
15 Ibid., 361–362.
16 Three-fifths majority was needed to amend the constitution and to elect the President, in accordance with the original constitutional provision.
17 Ibid., 364.
18 Wolchik (1997): 231.
19 Butora and Butorova (1999): 97.
20 Fish (1999): 49.
21 Malova (2001): 375.
22 Ibid., 359.
23 For other examples see *East European Constitutional Review* 6, no. 2 and 3 (Spring and Summer 1997); *East European Constitutional Review* 8, no. 4 (Fall 1999); *East European Constitutional Review* 9, no. 1 and 2 (Winter and Spring 2000). I do not mean to impugn the Court or the validity of its decisions. Rather, I mention these cases to show that the rules of the game remained undetermined even by those assigned to ensure their clear observance.

24 *Nucleonics Week*, April 6, 1995: 10.
25 *Nucleonics Week*, March 3, 1995: 17.
26 *Nucleonics Week*, June 1, 1995: 4.
27 *Nucleonics Week*, March 7, 1996: 1.
28 Malova (2001): 353 gives several examples.
29 My thanks to Allen Lynch for this insight.
30 Fish (1999): 47.
31 Ibid., 51.
32 Ibid., 48.
33 Ibid., 49.
34 Ibid.
35 Both opposition and governing parties wanted most of all to obtain a three-fifths majority to force a constitutional amendment that would have rendered referenda unnecessary. Under Meciar, the method of power transfer established in the constitution had been challenged by the threat of referenda. Yet even the use of public input outside the election cycle remained unstable as both the President and the Constitutional Court continued to present unclear decisions regarding the legality of popular plebiscites. After opposition parties gained control of Parliament and the government, subsequent HZDS threats were troubling but not destabilizing due to their inability to approve the necessary legislation authorizing such a move. A petition drive to call for early elections in 2000 is an example of such a muted threat. The government opposed the move, the President petitioned the Constitutional Court, and the Court said it could not intervene. The petition ultimately failed.
36 Article 2.1 of the Slovak constitution.
37 Malova (2001): 367.
38 Ibid.
39 See *East European Constitutional Review* 7, no. 1 (Winter 1998) for more details.
40 The initial constitution stated the president would arise from Parliament, something later revisions changed to allow for the nation's first popular election, which took place in 1999.
41 Goldman (1999) gives extensive details of the rivalry and Meciar's actions in particular.
42 *Nucleonics Week*, January 22, 1998: 3.
43 Miroslav Lipar, in discussion with the author, September 2002.
44 The treaty explicitly calls for "adequate authority, competence and financial and human resources to fulfill it assigned responsibilities" (Convention on Nuclear Safety, 1994, Article 8.1). Those responsibilities include licensing, inspection and enforcement of safety requirements contained in domestic and international commitments (ibid., Articles 2.2, 7.2.i-iv). The CNS also calls for shutting down reactors that cannot be upgraded (ibid., Article 6).
45 *Nucleonics Week*, November 11, 1993: 1.
46 Anonymous Slovak official, in discussion with the author, September 2002.
47 Miroslav Lipar, in discussion with the author, September 2002.
48 Kracji personally ordered all the ballots with the original four questions dealing with NATO membership and presidential elections to be swapped for ones that only had the three questions dealing with NATO. See *East European Constitutional Review* 6, no. 2 and 3 (Spring and Summer 1997) for discussions of the larger processes.
49 Butora (1999): 98.
50 Opposition efforts to test the boundaries of Meciar's power were not as severe as the Prime Minister's actions to change the system in his favor. The fact that they did it at all gives credence to the focus on rules stability.
51 Western European Nuclear Regulators Association (2000). Another panel of Western safety experts, government officials, and industry representatives concluded the Slovak nuclear regulatory authority was strong.

52 The same argument can be made concerning coalition stability in the Dzurinda government. The Hungarian party regularly threatened to withdraw unless its agenda was met. This was a fact of life for the new government for several years. Neither coalition stability nor foreign policy alignments changed, as did the level of compliance. Only the level of stability in the rules of the game changed in a similar direction and magnitude.
53 Government of Slovakia (1999).
54 Imports from Western Europe as the European Union deregulated electricity markets made up most of the difference, with the rest coming from Russian natural gas supplies.
55 For some interesting insights on the process of learning to play by new rules see Koremenos (2001); McCauley (1989); Kelman (1958).
56 While it is also possible to ascribe increasing certainty to a change in government alone, that change brought with it more than a pro-Western policy shift. The new government also emphasized the legal framework and democratic political behavior necessary for EU integration, not just piecemeal cooperation as under Meciar.
57 Bringing national legislation in-line with EU requirements had been the most obvious attempt. Minority rights, judicial independence, private property rights, and compliance with nuclear safety commitments showed the increasing harmony of elite behavior concerning political goals and how to achieve them.
58 Anonymous Slovak official, in discussion with the author, September 2002.
59 See European Commission (1999–2002) for further details.

References

Butora, Martin. "The Present State of Democracy in Slovakia." In *Democratization in Central and Eastern Europe*, edited by Mary Kaldor and Ivan Vejvoda, chapter 7. New York: Pinter, 1999.

Butora, Martin and Zora Butorova. "Slovakia's Democratic Awakening." *Journal of Democracy* 10, no. 1 (January 1999): 80–97.

International Atomic Energy Agency (IAEA). Convention on Nuclear Safety. 1994.

East European Constitutional Review 6, no. 2 and 3 (Spring and Summer 1997); 7, no. 1 (Winter 1998); 8, no. 4 (Fall 1999); 9, no. 1 and 2 (Winter and Spring 2000).

European Commission. *Annex to the Report on the Slovak Republic's Progress in Its Integration into the European Union during the Period of July 1–September 15 1999*. 1999a.

——. *Report on the Slovak Republic's Progress in Its Integration into the European Union, August 1998 to June 1999*. 1999b.

——. *Report on the Slovak Republic's Progress in Its Integration into the European Union, September 1999–June 2000*. 2000.

——. *Addendum to the Report on the Slovak Republic's Progress in Its Integration into the European Union, June–August 2002*. 2002a.

——. *Regular Report on Slovakia's Progress towards Accession*. 2002b.

Expert Mission of the IAEA Technical Cooperation Project. *Report of the Review of Results of the Gradual Upgrading at Bohunice VVER 440/230 NPP Units 1 and 2, November 19–24, 2000*. Document RER/9/046. 2000.

Fish, Steven. "The End of Meciarism," *East European Constitutional Review* 8, no. 1 and 2 (Winter and Spring 1999): 47–55.

Goldman, Minton. *Slovakia since Independence: A Struggle for Democracy*. Westport: Praeger, 1999.

Government of Slovakia, "Resolution 801/99 of the Slovak Government of 14th September 1999 and the EU Accession Protocol No. 9 on Unit 1 and 2 V1 Jaslovske Bohunice in Slovakia." 1999.

Kelman, Herbert. "Compliance, Identification, and Internalization: 3 processes of Attitude Change." *Journal of Conflict Resolution* (March 1958): 51–60.

Koremenos, Barbara. "Loosening the Ties that Bind: A Learning Model of Agreement Flexibility." *International Organization* 55, no. 2 (Spring 2001): 289–325.

Malova, Darina. "Slovakia: From the Ambiguous Constitution to the Dominance of Informal Rules." In *Democratic Consolidation in Eastern Europe: Institutional Engineering, Volume 1*, edited by Jan Zielonka, chapter 13. New York: Oxford University Press, 2001.

McCauley, Clark. "The Nature of Social Influence in Groupthink: Compliance and Internalization." *Journal of Personality and Social Psychology* 57 (1989): 250–260.

Nucleonics Week. August 5, 1993: 1; November 11, 1993: 1; September 8, 1994: 13; March 3, 1995: 17; April 6, 1995: 10; June 1, 1995: 4; March 7, 1996: 1; January 22, 1998: 3.

Western European Nuclear Regulators Association. *Nuclear Safety in EU Candidate Countries*. October 2000. Accessed multiple times 2000–2012. http://europa.eu.int/comm/dg1a/nss/index.htm.

Wolchik, Sharon. "Democratization and Political Participation in Slovakia." In *The Consolidation of Democracy in East-Central Europe*, edited by Karen Dawisha, and Bruce Parrott, chapter 6. New York: Cambridge University Press, 1997.

Zemko, Milan. "Domestic and International Aspects of the Czechoslovak State's Crisis and End." In *Irreconcilable Differences? Explaining Czechoslovakia's Dissolution*, edited by Michael Krause and Alison Stranger, Chapter 15. New York: Rowman and Littlefield, 2000.

Additional Reading

Energiewerke Nord GmbH and System Technology Management. *The Environmental Impact Assessment Report of V1 NPP Decommissioning, Non Technical Executive Summary*. 2006.

European Commission. *Resolution on Slovakia's Application for Membership of the European Union, with a View to the Vienna European Council*. December 11–12, 1998 – C 398/52.

Framework Agreement between the Slovak Republic and the European Bank for Reconstruction and Development Relating to the Activities of the Bohunice International Decommissioning Support Fund in Slovakia. 2001.

Grant Agreement between the Slovak Republic and the European Bank for Reconstruction and Development, Bohunice Nuclear Safety Project. 1993

Inter-parliamentary Union. "Slovakia Last Elections." Accessed multiple times 2000–2012. www.ipu.org/parline-e/reports/2285_E.htm.

Samson, Ivo. "Slovakia: Misreading the Western Message." In *Democratic Consolidation in Eastern Europe: International and Transnational Factors, Volume 2*, edited by Jan Zielonka and Alex Pravda, chapter 13. New York: Oxford University Press, 2001.

Wightman, Gordon. "Slovakia Ten Years after the Collapse of Communist Rule." In *Party Development and Democratic Change in Post-Communist Europe: The First Decade*, edited by Paul Lewis, chapter 8. Portland: Frank Cass, 2000.

7 Delaying disaster

> "It's easy to get disappointed about nuclear safety over there. At least Chernobyl closed, not that the Ukrainians wanted to do it though. We worked so hard to change safety culture, but unless they could see the light at the end of the tunnel they just wouldn't budge."
>
> —Anonymous US State Department official, June 2002

Nuclear power has several advantages over fossil fuels, not least its higher energy output per metric ton of fuel when compared in weight to oil or natural gas. Reactors can also operate in a broader range of geographical locations than sustainable energy sources like hydroelectric, wind, and thermal power.[1] The mining process poses environmental dangers, but so do deep sea oil rigs as seen in the 2010 Deep Horizon disaster. Nuclear power also raises concerns about long-term waste disposal, but it can still generate power more cheaply and with fewer airborne pollutants than coal-fired plants, the perennial bane of global environmental concerns. Newer reactor designs have also allowed for certain facilities to burn weapons-grade plutonium as part of post-Cold War disarmament programs, helping to reduce risks of nuclear smuggling and terrorism. Coupled with rapidly growing energy consumption in the developing world and demonization of fossil fuels for their role in global climate instability, it is easy to see why nuclear power will be around for the foreseeable future.

Safety technology has also vastly improved across the industry since the first-generation reactors. All modern plants have containment structures and extensive safety systems to monitor problems and prevent catastrophic accidents. Increasingly strict environmental standards also limit the daily radiation exposure for workers and surrounding areas, something especially important in the United States, Great Britain, and France since they have the highest concentration of nuclear power plants in the world. Other countries have not been so conscientious though, and the five in this study show the extent to which Western governments have gone to improve safety whenever possible, and demand early closure when not.

In contrast, Russia continues to operate unsafe RBMK reactors like Chernobyl, even recently extending their lives for another decade.[2] The Ministry of Atomic Energy included the oldest RBMK still operating in the world, Leningrad 1, in those extensions.[3] The need to export oil and gas supplies abroad drives this reliance on nuclear power in general, with more than sixty percent of the country's vast petroleum supplies going overseas to generate approximately fifty percent of federal revenues in the process.[4] Given those lucrative international markets and the concomitant political influence Russia has gained vis-à-vis the European Union and China in particular, the percentage of total energy produced by the country's RBMK's is projected to increase from seventeen percent currently to twenty-five percent over the next two decades to meet domestic energy needs without sacrificing the country's "cash cow."[5]

Worryingly, Russian design and day-to-day safety lags behind the West, and chances remain high that nuclear catastrophe will follow if the country experiences a major accident in the future; if Fukushima could cause so many lingering problems, how much more so reactors that lack its greater safety conditions. Despite assurances from Russian design engineers that the Chernobyl accident resulted from a single worker's incompetence, the tragedy demonstrated the fundamental flaws of any reactor lacking a containment structure. When the Russians and their national counterparts in the former Soviet bloc built their first-generation reactors they probably assumed an accident *could* not happen. Decades after Chernobyl, the Russians I spoke with said the same type of accident *would* not happen.

Safety upgrades had been planned for nearly all Russian nuclear reactors in the 1990s to make that confidence a reality, but work progressed slower than US negotiators hoped, with some aspects dragging into the new millennium. Delays resulted in part from the nuclear regulatory authority's dependence on the continued goodwill of the Ministry of Atomic Energy as it planned to expand nuclear power throughout the country, thereby making safety upgrades less desirable when they required temporary shutdowns, as the most serious ones did. The other part of the problem came from the risk acceptant safety culture and strong sense of national pride driving Russia's nuclear sector. In addition, while elites eagerly accepted Western money to clean up Soviet-era messes, often the money did not find its way into safety upgrades as promised, and reactor safety suffered unnecessarily for years.

Despite the delays though, operational safety has improved since the Chernobyl disaster, but unfortunately the inherent potential for catastrophic contamination of the Russian people remains unchanged given the lack of containment structures on the eleven RBMK reactors currently operating in the country. Adding to those potential woes, dozens of submarine reactors lay rotting for years at the Murmansk docks awaiting disposal in Russia's northernmost inhabited region.[6] For a time even the ground at the Kola nuclear facility became so contaminated by radioactive waste that workers attempted to walk above the festering mud using wooden blank bridges.[7] Sadly, such paltry precautions did nothing to prevent the contamination already destroying their lives and poisoning the surrounding arctic ecosystem; a growing mountain of solid waste also remains a potent threat even with plans to dispose of it in a modern storage facility.[8]

Yet despite the myriad problems, no Russian government ever entered into early closure commitments, thereby making comparative compliance behavior impossible, even though the need to evaluate the Russian nuclear industry will

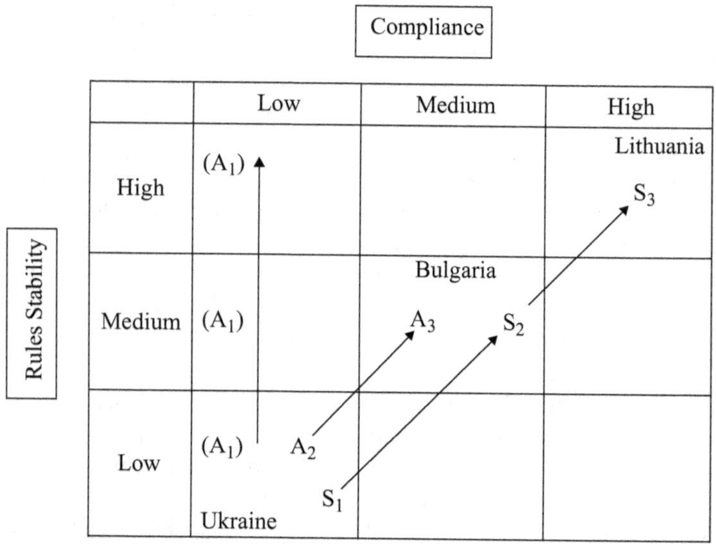

Figure 7.1 Rules Stability and Compliance

remain a priority of global safety concerns for decades to come. Thankfully, the five countries in this book followed different paths, some successfully achieving comparable safety to Western standards, others closing their dangerous reactors, with only one remaining obstinately open. The rules of the game defined how elites evaluated the costs of those actions, shaping elite expectations about the durability of the political systems and their individual participation in them. Undefined and unpredictable rules led to short-term thinking and compliance suffered. In contrast, stable systems allowed elites to see that the long-term benefits of compliance far outweighed the immediate costs, large as they seemed while their countries struggled to rebuild shattered economies and restore confidence among populations eager for change, but often not patient enough to wait for it. Conflicting aspects of stability produced compliance behavior that neither fully rejected agreement requirements, nor achieved the highest goals of closing unsafe reactors on time, if at all. Figure 7.1 and Table 7.1 illustrate the outcomes and trajectories among the case studies.

These five cases show more than positive correlation between rules stability and compliance. They confirm the causality of rules effects on elite time horizons concerning the costs and benefits of compliance. Taken together, the cases also show internal variation and variation across countries with comparable background conditions. While the causes for rules stability differed in places and over time, their effects remained consistent throughout all of the countries studied here. As a result, Slovakia and Lithuania both achieved high compliance, and while initially Armenia looked like it could get there as well, the country fell back into the same kind of

Table 7.1 Formal and Informal Rules Stability

Country	*Rules Stability*							*Time Period*
(Time)	*Power Transfer*		*Delegated Authority*		*Absence of Challenges*		*Result*	
	Formal	*Informal*	*Formal*	*Informal*	*Formal*	*Informal*		
Ukraine	No	No	No	No	No	No	**LOW**	**1990s to Chernobyl closure December 2000**
Armenia$_1$	Yes	Yes	Yes	Yes	Yes	Yes	**Moving towards HIGH**	**Independence 1991 to Ter-Petrossyan's resignation in February 1998**
Armenia$_2$	No	No	No	No	No	No	**LOW**	**February 1998 to President Kocharian's resumption of full control in May 2000**
Armenia$_3$	Yes	No	Yes	No	Yes	No	**MEDIUM**	**May 2000 to Metsamor life extension 2007**
Lithuania	Yes	Yes	Yes	Yes	Yes	Yes	**HIGH**	**Independence 1991 to Decommissioning in 2009**
Bulgaria	Yes	No*	Yes	No*	Yes	No*	**MEDIUM**	**1990 to 2004 (1 and 2) and 2006 (3 and 4) closure of Kozloduy**
Slovakia$_1$	No	No	No	No	No	No	**LOW**	**Velvet Divorce in 1993 to HZDS defeat in September 1998**
Slovakia$_2$	Yes	No	Yes	No	Yes	No	**MEDIUM**	**September 1998 to Meciar's presidential defeat in May 1999**
Slovakia$_3$	Yes	Yes	Yes	Yes	Yes	Yes	**HIGH**	**May 1999 to 2008 closure of V1**

*Bulgaria had unstable informal rules for the majority of the time period; brief improvements enabled greater compliance and eventual closure, but still outside the agreed upon time frame unstable situation characterizing Ukraine. Thankfully, Armenia recovered enough at least to improve safety and ensure its continuance with a capable nuclear regulatory agency. As a result, Bulgaria and Armenia ended up at the same medium level, the former due to closing its reactors later than the original deadline, while the latter chose to stay there and not decommission Metsamor. As with the other countries, reliance on nuclear power did not drive the Armenian decision directly, rather the ways elites understood the positive and negative options facing them conditioned their choices. The unlikelihood of EU membership might have nudged Armenia closer to Russia with its risk acceptant nuclear safety culture, but Armenia still stood to gain a great deal more from cooperation the West, as evidenced when it committed to early closure in the Partnership and Cooperation Agreement and made progress toward keeping it. However, while dependence on nuclear energy and the influence of international factors remained consistent, compliance varied with the rules of the game in Armenia, as it did in the other countries as well.

Perhaps then, despite improvements to operational safety and the regulator's capacity to ensure they continue, Armenia is the one "failure" of Western compliance efforts: Metsamor continues to operate, and like its Russian counterparts, continues to threaten the global community with disaster if safety degrades, or worse, another "Chernobyl" occurs. While the Ukrainians would have added their troubled reactor to that list if they could have, thankfully we can all breathe a sigh of relief that the nuclear reactor that caused so much suffering now sits entombed. Assuming the final confinement and removal procedures continue as planned, it will never again threaten the world with disaster.

Yet even with the progress made in Eastern Europe and the former Soviet Union, Fukushima reminds us that nuclear power comes with a price, a price that safety measures can mitigate but never really do away with completely as it continues to spew toxic plumes and spread poisons across the Pacific. In that regard then, have we really drawn the correct lessons from closing the world's most dangerous reactors, or are we simply delaying disaster once again?

Notes

1 Concerns for seismic activity and potential instability limit their placements, as will increasing awareness of coastal vulnerabilities to tsunamis and hurricanes after Fukushima.
2 Crawford and Smith (2011).
3 *World Nuclear News*, December 2, 2013.
4 For 2012 figures see US Energy Information Administration (2013).
5 World Nuclear Association (2013).
6 Population density in Russia's other arctic regions is considerably lower.
7 Anonymous US State Department official, in discussions with the author, June 2002. See Jorn Madslien (2006) for more on the growing recognition of these problems.
8 Thomas Nilsen (2013).

References

Crawford, David and Rebecca Smith. "Russia to Extend Life of Aging Reactors." *Wall Street Journal*, September 22, 2011. Accessed November 1, 2013. http://online.wsj.com/news/articles/SB10001424053111904563904576584831235317\u200b42.

Madslien, Jorn. "Nuclear Waste Poses Arctic Threat." BBC News, October 19, 2006. Accessed March 1, 2010. http://news.bbc.co.uk/2/hi/europe/6058302.stm.

Nilsen, Thomas. "Giant Radioactive Waste Complex Taking Shape." *Barents Observer*, October 15, 2013. Accessed December 15, 2013. http://barentsobserver.com/en/security/2013/10/giant-radioactive-waste-complex-taking-shape-15-10.

US Energy Information Administration. "Russia." November 26, 2013. Accessed December 15, 2013. www.eia.gov/countries/cab.cfm?fips=rs.

World Nuclear Association. "Russia's Nuclear Fuel Cycle." Accessed December 15, 2013. www.world-nuclear.org/info/Country-Profiles/Countries-O-S/Russia–Nuclear-Fuel-Cycle/.

World Nuclear News. "Restored RBMK Back On Line." December 2, 2013. Accessed December 15, 2013. www.world-nuclear-news.org/RS-Restored-RBMK-back-on-line-0212137.html.

Appendix A
Map of Soviet-designed reactors

Appendix A: Map of Soviet Designed Reactors
Adapted from International Nuclear Safety, Pacific Northwest National Laboratory

Appendix B
Reactor locations, statuses, and designs

VVER-440/V-230 reactors

Country	*Reactors*	*First power*	*Status*
Armenia	Armenia 1	1976	Closed 1989
	Armenia 2	1980	Shutdown expected 2016
Bulgaria	Kozloduy 1 and 2	1974, 1975	Closed December 2002
	Kozloduy 3 and 4	1980, 1982	Closed December 2006
East Germany	Greifswald 1–4	1974, 1975, 1977, 1979	Closed 1990
Russia	Kola 1 and 2	1973, 1974	Shutdown expected 2018, 2019
	Novovoronezh 3 and 4	1971, 1972	Shutdown expected 2016, 2017
Slovakia	Bohunice V1 units 1 and 2	1978, 1980	Closed December 2006, December 2008

RBMK reactors

Country	*Reactors*	*First Power*	*Status*
Lithuania	Ignalina 1	1983	Closed December 2004
	Ignalina 2	1987	Closed December 2009
Russia	Kursk 1	1976	Operating until 2021
	Kursk 2	1979	Operating until 2024
	Kursk 3	1984	Operating until 2024
	Kursk 4	1986	Operating until 2016
	Kursk 5	–	Construction abandoned
	Leningrad 1	1973	Operating until 2018
	Leningrad 2	1975	Operating until 2020
	Leningrad 3	1979	Operating until 2024
	Leningrad 4	1981	Operating until 2025
	Smolensk 1	1983	Operating until 2022
	Smolensk 2	1985	Operating until 2025
	Smolensk 3	1990	Operating until 2020

Ukraine	Chernobyl 1	1977	Closed 1996
	Chernobyl 2	1978	Closed 1991
	Chernobyl 3	1981	Closed 2000
	Chernobyl 4	1983	Reactor destroyed April 1986
	Chernobyl 5	–	Construction cancelled
	Chernobyl 6	–	Construction cancelled

Adapted from World Nuclear Association

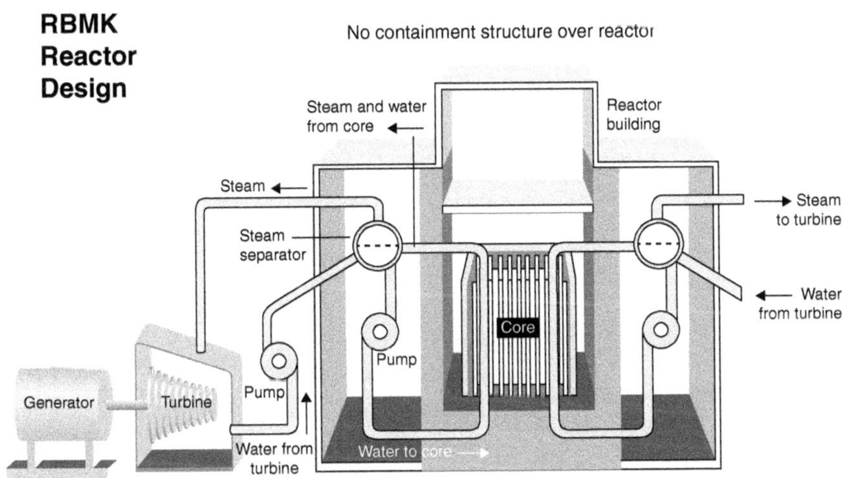

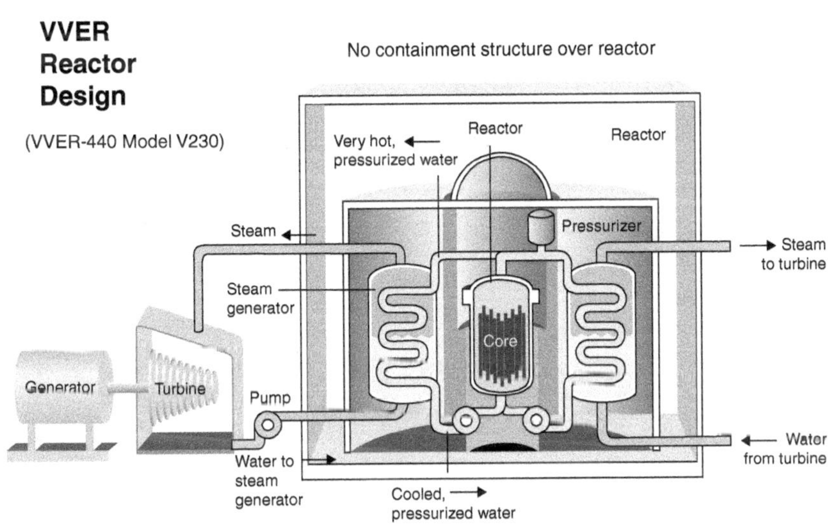

Appendix B: Reactor Locations, Statuses, and Designs
Adapted from Nuclear Energy Institute Source Book, www.nei.org/Master-Document-Folder/Backgrounders/Reports-And-Studies/Source-Book-Soviet-Designed-Nuclear-Power-Plants

Appendix C
Chernobyl Unit 4 after April 1986 disaster

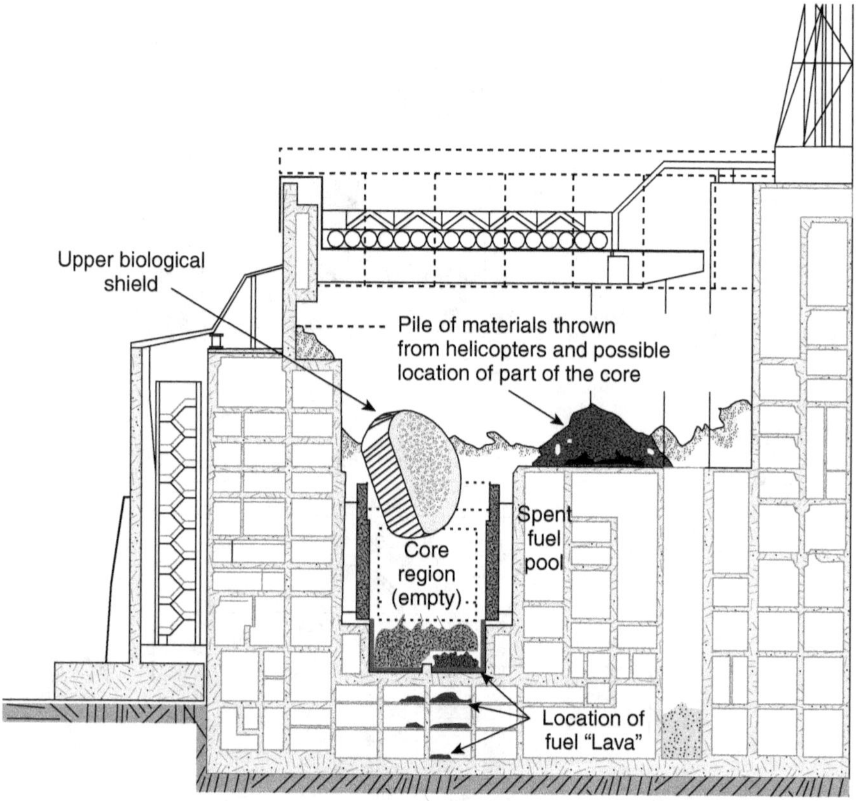

Appendix C: Chernobyl Unit 4 after April 1986 Disaster
Adapted from World Nuclear Association

Index

www.ingramcontent.com/pod-product-compliance
Lightning Source LLC
LaVergne TN
LVHW021037011026
849160LV00009BA/672
9781138282360